UNTAPPED

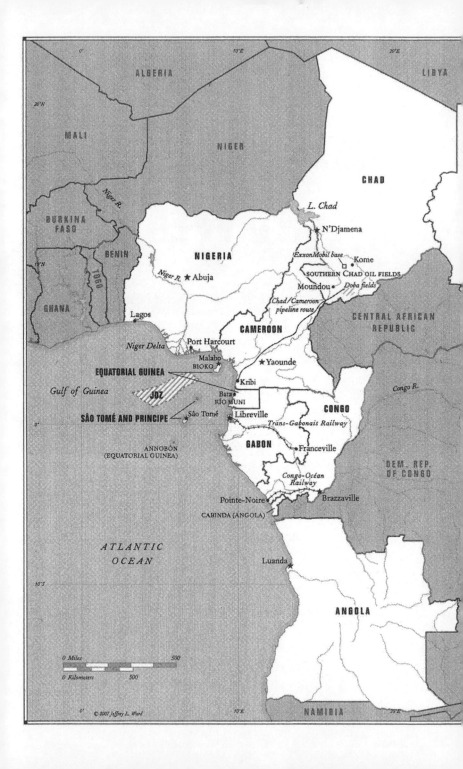

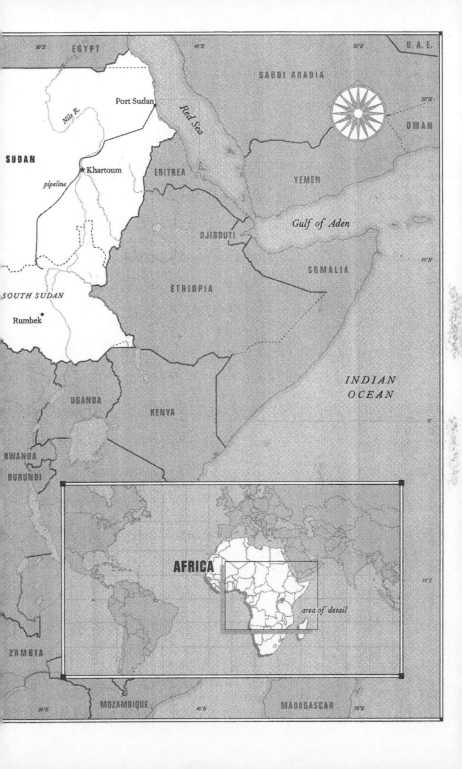

JOHN GHAZVINIAN

UNTAPPED

THE SCRAMBLE
FOR
AFRICA'S OIL

A HARVEST BOOK

HARCOURT, INC.

ORLANDO   AUSTIN   NEW YORK   SAN DIEGO   LONDON

The Library of Congress has cataloged the hardcover edition as follows:
Ghazvinian, John H. (John Hossein), 1974–
Untapped: the scramble for Africa's oil/John Ghazvinian.—1st ed.
p.   cm.
Includes index.
1. Petroleum industry and trade—Africa.   2. Petroleum—Africa.   I. Title.
HD9577.A2G43   2007
333.8'232096—dc22   2006030387
ISBN 978-0-15-101138-4
ISBN 978-0-15-603372-5 (pbk.)

Text set in Fournier MT
Designed by April Ward

Printed in the United States of America
First Harvest edition 2008
DOC 10 9 8 7 6 5 4 3

*To my father, who taught me
how the world worked,
and to my mother, who showed
me how to care.*

And in memory of Grandma.

CONTENTS

Preface

HER VOICE MADE ME think of Fargo.

It was one of those scrubbed-up, warbling voices from the Northern Plains, full of flattened vowels and Scandinavian resolve, and as it shuddered its way down the phone line, it seemed worth giving in to the image in my head. So I closed my eyes and thought of Fargo.

"When would you like to *trayvel*?"

If you closed your eyes, it was all there. The Peterbilt trucks and the speedboat auctions; the envelope of fresh November snow. The "family" restaurants feeding the great American stomach a steady diet of hometown pride and manky coleslaw. A faded gingham and rusty styrofoam version of America, full of all the contrived honesty of a twenty-cent cup of truck-stop coffee and an evening with the Weather Channel. A postcard from the Great White North.

Okay, so I got a little carried away.

It probably wasn't Fargo on the line. Maybe it was Duluth, or Saginaw or Grand Forks or Fond du Lac. Or even suburban Delaware. Not like I'd know the difference. In fact, there's a good chance I was talking to a call center in Bangalore. But if you have to book airline tickets by phone, why not daydream a little?

It was November 2004, and we were living in a world where you could make a video clip of yourself and send it to a BlackBerry in Tora Bora in less time than it was taking to spell my last name to this woman on the phone. I couldn't remember the last time I bought airline tickets without using the Internet, and the whole thing felt a little odd, a little inefficient—like something your parents might do. But I was flying to Nigeria, and if you want to fly to Nigeria, you have to buy your ticket the old-fashioned way. Even if you found the fare online, you have to book your seat over the phone and then go down to the airport to pay for it. They literally have to see the credit card in your hand.

There is a good reason for this low-tech red tape. During the 1990s, under the military dictatorship of General Sani Abacha, Nigeria became an internationally acknowledged entrepôt for money laundering, narco-trafficking, and organized crime. Even now, scarcely a nanosecond goes by without someone, somewhere in the world receiving an e-mail from someone claiming to be the country's foreign minister, in desperate need of a sort code and an account number and an electronic pen pal who can help him park a sum of $25 million. They call the scam a "419" in Nigeria, after the relevant penal code in Nigerian law, and the practice has added a high-tech mystique to the country's unsavory reputation. The international banking industry remains institutionally paranoid about any transaction that involves Nigeria, which means, among other things, that you can forget about buying tickets to Lagos over Travelocity.

Driving down to LAX just to show my face and good faith was only my first glimpse of the many hassles and indignities of life in a country with a dysfunctional commercial-banking system. Once in Nigeria, I would see people paying for real estate with enormous sacks of cash that would take hours to count, thanks to years of runaway inflation. Just imagine—paying for a house with $5 bills. And those were the lucky ones. People who could afford houses.

But for the moment, all that lay in the future. For the moment, I was on the phone to Fargo.

"So what is it that's taking you over there, anyway?" she asked while we were waiting for one of her screens to come up. "Business or pleasure?"

"Business, I suppose," I said. "I'm doing some research for a book."

"Oh yeah? What's the book about?"

"Well, it's about oil. Oil in Africa."

"Oh, they got oil in Africa?"

"Yes, there's quite a lot of it," I replied, pleased with myself for drumming up a little advance publicity in the American heartland. "And we're starting to get more and more of our oil from over there." I was just getting warmed up. "In fact, Nigeria has been—"

"Good!" she said, with a burst of indignation totally out of character for a customer service agent. "We have to get it from somewhere."

"Well, um, sure, but of course it's not always that simple," I said, clearing my throat. It felt like a bizarre reversal of roles, me trying to soothe this prickly customer. In any case, it turned out to be the wrong thing to say. A sheet of early-evening ice made its way quietly across Fargo.

"Window or aisle?"

◄◄-►►

We have to get it from *somewhere*.

That just about says it all, doesn't it? That is the American condition in the twenty-first century, and it's hard to argue with. The Arabs have let us down, the environmentalists won't let us drill in Alaska, and even dear little Venezuela's getting cocky. So what are we to do? Public transport? Have you seen the *size* of this country?

Outside that call center in Fargo, there will be a parking lot. It will be a checkerboard of Buick LeSabres and Chevy Caprices and light trucks and, yes, even an SUV or two. None of the people strapped into cubicles and headsets all day will have another way of getting home after they're done booking people onto flights to Lagos. There is no bus they can hop on, as there would be almost anywhere else in the developed world. America just wasn't built that way.

And it isn't simply a question of SUVs or Hummers or miles per gallon, despite what America's bumper-sticker playground of political discourse might have you believe. I have spent my childhood in London, my adolescence in the United States, and much of my adult life back in Britain. Every time I return, America's relationship to energy looks more bizarre. You don't have to be an alfalfa-munching hippie living in a tree house to notice that here in the United States, we go through oil like it's water.

But there is no point getting sanctimonious about it. America is not going to change into Europe overnight. We are not going to wake up one morning happy to live as a nation of miniature fridges, under-heated homes, lukewarm soft drinks, and clothes hung out to air-dry. This is something American conservationists, like much of the world, will have to come to terms with for the time being. What makes us American is that we take our personal comfort seriously, and treat its instant realization as a sort of extreme sport. The corollary to that, and our collective Achilles' heel as a nation—as we all know—is the amount of nonrenewable energy it takes to maintain that way of life.

So, yes, Fargo was right. Until someone comes up with a better idea, we have to get it from *somewhere*.

◄◄•►►

This book is a journey into that somewhere. It is a journey into a part of the world that for most of us never transcends the familiar images of sweltering, fly-infested refugee camps; starving, wide-eyed waifs; and truckloads of doped-up child soldiers speeding through dusty villages.

In our hearts, we want to believe that "Africa" is so much more than this never-ending diorama of despair and human suffering, this biblical landscape of plagues and famines and smiting armies. And every now and again, when we learn to look for them, we see signs of promise. Every now and again, Africa seems ready to turn the corner, ready to embrace those who are drawn by the centuries-old allure of its considerable natural wealth. Every now and again, we are told to look past the perpetual slide show of maggots and blood and bloated corpses, and to hear the message that Africa is open for business.

We are living through another one of those now-and-agains. Thanks to more than a decade of wildly successful discoveries by the world's largest oil companies, as well as the efforts of a growing army of Washington lobbyists and lawmakers, Africa has been quietly transformed in policy-making circles from an insignificant backwater into a potentially lucrative new source of oil and gas for the global market. (Listen to some of the more zealous advocates and you may even hear wild talk about how it may soon "replace the Middle East.") This book is the story of that transformation, and an attempt to understand what it might mean—for Africa, for America, and for the world.

It is also the product of a fascination with Africa that dates back nearly twenty years. It was in a high-school library in suburban Los Angeles in the late 1980s, when I was fifteen years old, that I first read about a tiny tropical country called Equatorial Guinea, where the president had once lined up his political opponents in a football stadium and had them gunned down to the accompaniment of rock music. It was the kind of story that sticks in your head, and I became something of an Equatorial Guinea junkie after that.

Back then, my Africa was still the Africa of flies and dust and child soldiers, and it never took much effort to follow the news from Equatorial Guinea, because there was hardly ever any. But a few years ago, I heard the country had discovered oil, and lots of it. American oil companies were investing billions and the Bush administration, said the press reports, was about to reopen the American embassy in the capital, Malabo.

I began to hear other things that were hard to ignore. Equatorial Guinea was just one of several African countries suddenly awash with oil money. Older producers, such as Nigeria and Angola, were rapidly increasing their output as well. I read that the United States might soon be getting as much as 25 percent of its imported oil from sub-Saharan Africa. I read that China was already heavily reliant on African crude, that international oil companies were investing billions in African exploration efforts, and that vast expanses of the African continent had never been properly explored for their hydrocarbon potential. With American and British forces still bogged down in a distant country whose strategic importance lay at least in part with its vast petroleum wealth, this information seemed to raise an interesting question. Shouldn't we all know a little bit more about where our oil will be coming from?

So, in 2005, I set out to see for myself what all the fuss was about. With a suitcase full of notepads and malaria pills, and a sweaty money belt stuffed with $100 bills, I spent six months traveling through twelve African* countries—from Sudan to Congo to Angola, and just about everywhere in between—hoping to hear a little bit more about the challenges, the obstacles, the reasons for hope and the reasons for despair. In some countries I visited, such as Nigeria, vast amounts of oil had been flowing for decades and the industry was a major player in the nation's politics and economy. In others, such as São Tomé, no one had even drilled a hole yet, but there was excited talk about the petrodollars on the horizon. I talked to politicians, economists, warlords, diplomats, aid workers, oil-company executives, local journalists, activists, priests, political prisoners, crude-oil bandits, cabdrivers, soldiers, missionaries, bureaucrats, technocrats, scientists, rebel-militia leaders, historians, oil-rig workers, lawyers, bankers, and even a few children and muttering old men, just for good measure.

---

*Students of North Africa will quickly notice that this book, like so many others purporting to be about "Africa," focuses almost entirely on countries south of the Sahara and will perhaps query the decision, given that Egypt and Algeria are both significant oil and gas producers and Libya is rapidly increasing its output. But while the nations of the Maghreb are no less "African" than those south of the Sahara, my goal here has been to avoid the far more familiar terrain of Arab (and, by extension, North African) oil politics. Some will consider this a criminal omission, but any attempt to shoehorn in a superficial discussion of these countries in the absence of a larger exposition of Arab history and politics would, I would argue, have been the greater crime.

What I came back with was a suitcase bulging with notes and a head full of questions. To try to make sense of it all, I spent a few more months nosing around Washington, London, and Paris, talking to the phalanxes of analysts, lobbyists, and academics who make Africa their bread and butter. What I ended up with was a nagging feeling that we are ignoring this story at our peril.

In 2001 Tony Blair described "the state of Africa" as a "scar on the conscience of the world." Later that year, George W. Bush referred to Africa as a disease-ridden "nation." Each leader unwittingly revealed the prevailing attitude of his people to Africa—the one full of facile posturing and lingering postcolonial guilt, and the other, well, let's be charitable and call it homespun. But what both men also demonstrated was just how difficult it can be to talk about "Africa"— a continent of 54 countries, 2,000 ethnic groups, and 3,000 languages— without lapsing into sanctimoniousness or demonstrating frightening levels of ignorance. What follows is a snapshot of a moment in time when Africa seems on the verge of playing a bigger role in world energy security, and the beginnings of a conversation about the complexities and challenges that form the backdrop to one of every nine gallons of fuel being pumped into any given car, anywhere in the world, on any given day.

Now, the real question is: Will it play in Fargo?

UNTAPPED

Introduction

As bright colors that denote and dominate the heavens,
We CHARGE the clouds!

The winds! The rains!

We ensure the fertility of Mama Afrikah!

IT WAS EARLY in the evening, still the hour of terrines and spritzers and stiff handshakes. But for Chigomezgo Miriam Gondwe, an irrepressibly sensual young spoken-word performer from Malawi, there was no reason to hold back.

In her womb, hopes are raised to create opportunities!

The time is ideal for us to emerge
To view with these once-tortured eyes
The delight on her anxious and excited face!

Gondwe, who describes herself as an "ethno-urban hip-hop soul poetess," was clothed in a sunflower-yellow traditional African dress and head wrap, and was spitting sizzle and brimstone, with a look of raw, sexual ecstasy on her face not unbecoming of someone seized by the Rapture. She had appeared out of nowhere on the stage at the Coca-Cola Dome in the Johannesburg suburb of North Riding, and now she was spiritedly reciting some of her work in front of around 3,500 assembled conference delegates.

> *All of our stories, individual and collective*
> *Are born of the African dawn!*
>
> *So continue to strive, mah Aff-ree-KAH!*
> *And hold on to your pride*
> *And Africa's sun shall never set on your greatness*
> *Mah Aff-ree-KAH*
> *Africa's sun shall NOT set!*

To their considerable credit, the ocean of mostly gray, suited, mustachioed men from every corner of the globe were trying hard not to look disconcerted by this unexpectedly fierce display of sistah-hood.

It was (and could only have been) the official opening ceremony of the 18th World Petroleum Congress, a massive working pow-wow and lavish spectacle held once every three years and generally described as "the Olympic Games of the petroleum industry." The moniker is an apt one, as, when it comes to sheer pomp and pageantry, there is nothing else quite like it in the world of hydrocarbons and petrochemicals. At the 17th Congress, held in 2002 in Rio, delegates had partied to samba music late into the night, as their Brazilian hosts passed the ceremonial flag to South Africa. Since it takes in all aspects of the oil business, from global politics to finance to geophysics, the WPC has evolved into far and away the most important date on the industry's calendar, a rare opportunity for Gulf sheikhs in flowing robes to dine and dance with Venezuelan socialists in between discussions of lateral-drilling technology, international fiscal-compliance regimes, and the Caspian Miocene shelf. Possibly the only thing missing from the festivities is a sandaled runner carrying a perpetual flame.

In the years since the first Congress was held in London in 1933, the choice of venue has often acted as a reflection of profound changes within the international petroleum industry. When the WPC was created, motorcars were seen as toys for the mega-wealthy, coal was still king, and oil exploration felt like a cutting-edge industry poised to supply the fuel of the future. No one had heard of "peak oil" or OPEC or Hugo Chavez, and most of today's oil-rich nations were still colonies and protectorates administered by Britain and France. BP, or British Petroleum, was still known as Anglo-Iranian Oil Company. The world of oil, like virtually everything else, revolved around London. But in the decades that followed, the WPC would hold court in such upstart industry hubs as Mexico City (1967), Houston (1987), and Calgary (2000).

In September 2005 the traveling circus had come to Johannesburg, the first time in its seventy-two-year history that the WPC had convened in Africa. The decision to do so was being touted as a nod to the continent's growing importance as an oil-producing region, and the savvy South African hosts were not about to overlook an opportunity to milk the Afrikah angle for everything it was worth.

The evening had begun with delegates being dropped off by the busload in front of the Dome. We had sashayed along the red carpet as a grinning Soweto steel band played their hearts out from just behind the velvet rope. Inside, another traditional African band played on stage as delegates searched for their tables. Fifty-foot banners hung from the rafters, with messages like "African Dawn—World Energy Solutions Created in Africa" and "We welcome you with African pride to the 18th World Petroleum Congress." A gigantic disk marked with a PetroSA logo was suspended from the ceiling, reminder that the South African oil company would be picking up the tab for the evening's festivities.

As the first course arrived—a trio of terrines made with smoked salmon, Kabeljou, smoked snoek, and angelfish—the interior of the Dome went pitch-black. Two enormous video screens suspended over the stage came to life, showing a dramatic montage of classic African scenes, set to a stirring, primordial drum-driven sound track that could only be described as Afro-electronica. "THE PULSE . . . OF AFRICA!" bellowed a baritone Voice of God, with a reverberation that made the terrines wobble. Sprawling savannahs, lush forests, and hissing leopards flashed by on the screen as the music played on and the Voice continued to rumble, backed up by the occasional thumping of kettledrums. Glistening fish leaped out of virgin rivers, women beamed over their meager wares at rustic markets, Tuareg tribesmen beckoned across rippling sand dunes. Children laughed as they pulled wooden wheelie toys on strings. It was like a prelude to Armageddon scripted by *National Geographic.*

Sweating, shirtless steel drummers now appeared on spotlighted podiums next to each of our tables, gyrating and banging away on their drums with a limber adolescent energy. Images of oil refineries and deepwater drilling platforms jumped from the screen as runners with flowing kitelike banners began sprinting between our tables. The Voice began reciting a poem about wind and dreams.

"THE TIME HAS COME TO LET OUR LIGHT SHINE," boomed the Voice. More half-naked men appeared, this time performing a warrior dance. The other journalists at my table munched away at their terrines, not about to let any impending apocalypse get in the way of a free meal. "BEHOLD THE RISING OF A NEW DAWN— AN *AFRICAN* DAWN—WITH RAYS OF LIGHT HERALDING ENERGY'S FUTURE," thundered the Voice. "ENERGY . . . IS IN THE BEAUTY AND POWER THAT IS AFRICA. THE WEALTH AND RESOURCES IN THE EARTH BENEATH OUR FEET." This was clearly a reference to Africa's oil wealth, but to keep things subtle, cheetahs leaped across verdant grasslands and scooters sped down bustling city streets on the video screens.

"THIS IS OUR TIME. THIS . . . IS OUR . . . AFRICA," the Voice proclaimed, as the music and the video montage built to a climax. I looked to the stage, half-expecting a cloud of smoke and the arrival of an African messiah, flanked by live panthers. Instead, the spotlight went up on a band of singing women draped in South

African flags. Then Chigo Gondwe came onstage and began reciting her Afro-positive performance poetry, backed up by a little gentle ululating from the women. After a few more poems, and a little more ululating, the opening medley drew to a close as fireworks shot out from the front of the stage.

It was a tough act to follow, but Desiderio Costa, the Angolan petroleum minister, didn't do himself any favors with his poor, halting English as he read from a prepared speech he clearly hadn't written, or even rehearsed. Angola was a cosponsor—along with Nigeria, Libya, and Algeria (Africa's four biggest oil and gas producers)—of the 18th WPC, and Costa's speech was followed by equally anodyne performances by officials from the other three countries. By the time the Libyan took to the stage, most delegates were slumped in their seats and had settled into idle chitchat about the chicken in saffron basil sauce that sat in front of them.

Salvation came in the form of Thabo Mbeki, South Africa's dynamic president, whose role was to declare the WPC officially open. Mbeki gave an eloquent speech about the dangers of Afro-pessimism, drawing on Duke Ellington and W. B. Yeats, and warning against the complacency of what he called "mood indigo," before slipping off, by his own admission, to a jazz club on the other side of Johannesburg. A special performance by the South African sensation Umoja rounded out the evening's festivities, and the only thing left was to bring out the desserts.

And it has to be said that the evening would not have been the same without the desserts. The organizers had decided to give us each a little chocolate mousse and sponge cake carefully molded into the shape of Africa. It was hard not to admire the culinary artistry involved, but as I looked round the Dome, I wondered: was I the only one to pick up on the symbolism of 3,500 drunken oil executives devouring the Dark Continent, bite after dribbling, chocolaty bite?

The mood of manufactured Afro-positivity was fresh in the air the next morning as the Congress got under way in Sandton City, an upmarket business park, hotel, and shopping center in Johannesburg's wealthy northern suburbs. The opening plenary session was called "The African Perspective," and made it clear, in case anyone had missed the point the night before, that Africa was ready for the international oil industry's embrace.

And as if this wasn't enough, on the third night of the Congress we were all invited to Gold Reef City, a gold-rush heritage theme park on the south side of Johannesburg, for "Africa Night," sponsored by South African Airways. SAA had rented out the park for the night, and laid on dozens of traditional performers from all corners of Africa, along with traditional foods representing each of Africa's oil-producing countries—Nigerian jollof rice, Angolan *calulu de peixe*, and so on—all topped off with a giant Lindt chocolate fountain for dipping desserts.

By the end of the five-day Congress, only a red-assed baboon could have failed to appreciate the take-home message to the international oil industry. *Africa: Come and get it.*

◄◄—►►

As if they ever really needed an invitation.

The WPC could have skipped the jollof rice and the steel bands and the live fireworks. These days, herds of stampeding wildebeest couldn't keep the international oil industry away from Africa. Since the early 1990s, advances in deepwater-drilling technology and attractive contractual terms have helped turn Africa into the world's last true El Dorado—a place where exploration blocks the size of France can still be picked up at an auction, and host governments lack either the experience or the technical capacity to impose burdensome constraints on drilling activity. For years Africa suffered from its image as a bad place to do business—racked by instability, corruption, and political violence—and in many ways, it still does. But as the world has begun to run out of big new oil bonanzas, the industry's appetite for risk has grown considerably.

You can see it in on board the fleet of MD-11s that leave the tarmac in Houston a couple times a week, bound for destinations with names exotic to the Texan ear, like Luanda and Malabo. Operated by World Airways and nicknamed the "Houston Express," the nonstop flights are open only to members of something called the US–Africa Energy Association, and seats start at $5,915 for business class. They rarely go empty.

You can see it, too, in Paris, where Air France runs its nonstop, private "Dedicate" service to an ever-growing number of African oil cities and encourages regular travelers to join its "Petroleum Club" to take advantage of "exclusive services for the oil and gas industry."

And you could see it in Johannesburg at the WPC, where the country presentations from Nigeria and Angola were so heaving with spectators that most delegates were reduced to peering in from the hallway outside.

Since 1990 alone, the petroleum industry has invested more than $20 billion in exploration and production activity in Africa. A further $50 billion will be spent between now and the end of the decade, the largest investment in the continent's history—and around one-third of it will come from the United States. Three of the world's largest oil companies—the British-Dutch consortium Shell, France's Total, and America's Chevron—are spending 15 percent, 30 percent, and 35 percent respectively of their global exploration and production budgets in Africa. Chevron alone is in the process of rolling out $20 billion in African projects over a five-year period.

The overwhelming majority of this new drilling activity has taken place in the so-called "deep water" and the "ultradeep" of the Gulf of Guinea, the roughly 90-degree bend along the west coast of Africa that can best be visualized as the continent's "armpit." Its littoral zone passes through the territorial waters of a dozen countries, from Ivory Coast in the northwest down to Angola in the south, and a good deal of its geology shares the characteristics that have made Nigeria a prolific producer for decades. Indeed, a number of unexpectedly productive fields have been discovered in the Gulf over the past decade. But although the Gulf of Guinea has lately been sub-Saharan Africa's most exciting region for the oil industry, it is hardly the only "prospective" part of the continent (to borrow the industry term). The parched semideserts of southern Chad and southern Sudan have recently added hundreds of thousands of barrels a day to global markets, and a growing chorus of voices is now touting the East African margin as the industry's "next big thing."

But be it east or west, jungle or desert, it is a safe bet that where the drillers go, the politicians, strategists, and lobbyists are not far behind. Washington in particular has taken a keen interest in Africa's growing significance as an oil-producing region since the headline discoveries of the late 1990s. In December 2000 the National Intelligence Council, an internal CIA think tank, published a report in which it declared unambiguously that sub-Saharan Africa "will play an increasing role in global energy markets," and predicted that the region would provide 25 percent of North American oil imports by 2015, up from the 15 percent or so at the time. (This would put Africa well ahead of Saudi Arabia as a source of oil for the United States.) In May 2001 a controversial and fairly secretive energy task force put together by U.S. Vice President Dick Cheney declared in its report: "West Africa is expected to be one of the fastest-growing sources of oil and gas for the American market."

In the following months, a group of congressmen, lobbyists, and defense strategists came together under the umbrella of the African Oil Policy Initiative Group, and began preaching the message that the Gulf of Guinea was the new Persian Gulf, and that it should become a strategic priority for the United States, even to the point of requiring an expanded military presence. A series of well-placed articles in the American media followed, some breathlessly announcing the inauguration of a new Middle East off the shores of Africa. Before long, the influential Center for Strategic and International Studies had chimed in with a couple of reports, its most recent, in July 2005, claiming that "an exceptional mix of U.S. interests is at play in West Africa's Gulf of Guinea."

During these years, a number of prominent lawmakers in Washington began getting excited about the possibility of shifting some of America's oil dependence from the Middle East to Africa. One former senior official charged with African affairs recalls Kansas Senator Sam Brownback rushing up to him one afternoon in October 2002, positively glowing with excitement. "What do you think about bases in Africa?" Brownback asked. "Wouldn't that be great?"

-<--->-

But does Africa measure up to the hype? After all, the entire continent is believed to contain, at best, 10 percent of the world's proven oil reserves, making it a minnow swimming in an ocean of seasoned sharks. Africa is unlikely ever to "replace" the Middle East or any other major oil-producing region. So why the song and dance? Why all the goose bumps? Why do so many influential people in Washington let themselves get so carried away when they talk about African oil?

The answer has very little to do with geology. Africa's significance as an oil "play," to borrow the industry lingo, lies beyond the number of barrels that may or may not be buried under its cretaceous rock. Instead, what makes the African oil boom interesting to energy-security strategists in both Washington and Europe (and, increasingly, Beijing) is a series of serendipitous and unrelated factors that, together, tell a story of unfolding opportunity.

To begin with, one of the more attractive attributes of Africa's oil boom is the quality of the oil itself. The variety of crude found in the Gulf of Guinea is known in industry parlance as "light" and "sweet," meaning it is viscous and low in sulfur, and therefore easier and cheaper to refine than, say, Middle Eastern crude, which tends to be lacking in lower hydrocarbons and is therefore very "sticky." This is particularly appealing to American and European refineries, which have to contend with strict environmental regulations that make it difficult to refine heavier and sourer varieties of crude without running up costs that make the entire proposition worthless.

Then there is the geographic accident of Africa's being almost entirely surrounded by water, which significantly cuts transport-related costs and risks. The Gulf of Guinea, in particular, is well positioned to allow speedy transport to the major trading ports of Europe and North America. Existing sea-lanes can be used for quick, cheap delivery, so there is no need to worry about the Suez Canal, for instance, or to build expensive pipelines through unpredictable countries. This may seem a minor point, until you look at Central Asia, where the Baku-Tbilisi-Ceyhan pipeline, stretching from Azerbaijan through Georgia and into Turkey, and intended to deliver Caspian crude into the Mediterranean, had to navigate a minefield of Middle East politics, antiglobalization protests, and red tape before it could be opened. African oil faces none of those issues. It is simply loaded onto a tanker at the point of production and begins its smooth, unmolested journey on the high seas, arriving just days later in Shreveport, Southampton, or Le Havre.

A third advantage, from the perspective of the oil companies, is that Africa offers a tremendously favorable contractual environment. Unlike in, say, Saudi Arabia, where the state-owned oil company Saudi Aramco has a monopoly on the exploration, production, and distribution of the country's crude oil, most sub-Saharan African countries operate on the basis of so-called production-sharing agreements, or PSAs. In these arrangements, a foreign oil company is awarded a license to look for petroleum on the condition that it assume the up-front costs of exploration and production. If oil is discovered in that block, the oil company will share the revenues with the host government, but only *after* its initial costs have been recouped. PSAs are generally offered to impoverished countries that would never be able to amass either the technical expertise or the billions in capital investment required to drill for oil themselves. For the oil company, a relatively small up-front investment can quickly turn into untold billions in profits.

Yet another strategic benefit, particularly from the perspective of American politicians, is that, until recently, with the exception of Nigeria, none of the oil-producing countries of sub-Saharan Africa had belonged to the Organization of Petroleum Exporting Countries (OPEC).* Thus they have not been subject to the strict limits on output OPEC imposes on its members in an attempt to keep the price of oil artificially high. The more non-OPEC oil that comes onto the global market, the more difficult it becomes for OPEC countries to sell their crude at high prices, and the lower the overall price of oil. Put more simply, if new reserves are discovered in Venezuela, they have very little effect on the price of oil because Venezuela's OPEC commitments will not allow it to increase its output very much. But if new reserves are discovered in Gabon, it means more cheap oil for everybody.

But probably the most attractive of all the attributes of Africa's oil boom, for Western governments and oil companies alike, is that virtually all the big discoveries of recent years have been made off-shore, in deepwater reserves that are often many miles from populated land. This means that even if a civil war or violent insurrection breaks out onshore (always a concern in Africa), the oil companies can continue to pump out oil with little likelihood of sabotage, banditry, or nationalist fervor getting in the way. Given the hundreds of thousands of barrels of Nigerian crude that are lost every year as a result of fighting, community protests, and organized crime, this is something the industry gets rather excited about.

Finally, there is the sheer speed of growth in African oil production, and the fact that Africa is one of the world's last underexplored regions. In a world used to hearing that there are no more big oil discoveries out there, and few truly untapped reserves to look forward to,

---

*In January 2007, Angola became the first new member of OPEC in over thirty years, and Sudan is expected to join late in 2007. Gabon withdrew from the organization in 1995.

the ferocious pace and scale of Africa's oil boom has proved a bracing tonic. One-third of the world's new oil discoveries since the year 2000 have taken place in Africa. Of the 8 billion barrels of new oil reserves discovered in 2001, 7 billion were found there. In the years between 2005 and 2010, 20 percent of the world's new production capacity is expected to come from Africa. And there is now an almost contagious feeling in the oil industry that no one really knows just how much oil might be there, since no one's ever really bothered to check.

All these factors add up to a convincing value proposition: African oil is cheaper, safer, and more accessible than its competitors, and there seems to be more of it every day. And, though Africa may not be able to compete with the Persian Gulf at the level of proven reserves, it has just enough up its sleeve to make it a potential "swing" region—an oil province that can kick in just enough production to keep markets calm when supplies elsewhere in the world are unpredictable. Diversification of the oil supply has been a goal—even an obsession—in the United States since the Arab oil embargo of the 1970s. Successive U.S. administrations have understood that if the world is overly reliant on two or three hot spots for its energy security, there is a greater risk of supply disruptions and price volatility. And for obvious reasons, the effort to distribute America's energy-security portfolio across multiple nodes has taken on a new urgency since September 11, 2001. In his State of the Union address in January 2006, President Bush said he wanted to reduce America's dependence on Middle East crude by 75 percent by 2025.

◂◂‑▸▸

Some of the more evangelical proponents of African oil have argued that here, at last, is the longed-for "clean break"—the chance to detach the fortunes of America once and for all from Middle East crude. For several decades, the United States and other Western governments made controversial compromises with despotic rulers across the Middle East, from the Shah of Iran to the House of Saud, in an effort to keep the oil flowing. Western petrodollars turned nomadic tribes into wealthy emirates, with scant concern for democracy or human rights. American support for undemocratic and unaccountable governments bred great resentment across the region, the consequences of which the entire world is now coming to grips with. This time, in Africa, these advocates say, we have a chance to start fresh and "get it right."

But how do we know African oil really represents a good-bye to all that? In many ways the situation in Africa is far more difficult, complex, and dangerous than the situation in the Middle East. Africa is filled with so-called "failed" states, or states that teeter perpetually on the edge of failure, one disputed election away from outright conflagration. Illicit arms are traded across fluid, largely fictional borders, national sovereignty is little more than a collection of flags and anthems, ethnic tribalism is alive and well, and angry militias have already turned to stealing crude oil as a way to keep themselves in business. According to the National Maritime Institute, the Gulf of Guinea is the world's second-most-dangerous waterway.

Some feel that in searching for an alternative to the volatile politics of the Middle East, Washington strategists have focused on a constellation of far more troubled and impoverished nations, from Angola to São Tomé. According to this narrative, as the international media spotlight has stayed trained on Iraq and the Middle East, an unholy alliance has quietly formed between think tanks, oil-industry lobbyists, PR firms, and entrepreneurial businessmen, all keen to rebrand and reposition corrupt and often violent regimes as benevolent and important new allies of the West. Human-rights campaigners warn that many of the fatal compromises that were made with undemocratic and unpopular rulers in the Middle East are being repeated all over Africa, with potentially catastrophic consequences. In the worst cases, control of oil revenue has sparked violent conflicts, with already-bitter divisions exacerbated by the promise of untold riches for the victor.

But it is not just the usual amen chorus of humanitarian Chicken Littles and louche, *bien pensant* urban intellectuals who have sounded alarm bells over the African oil boom. Hard-nosed economists, too, are skeptical. A growing body of evidence suggests that oil, far from being a blessing to African countries, is a curse. Without exception, every developing country where oil has been discovered has seen its standard of living decline and its people suffer, while its less-well-endowed neighbors have gone on to (relative) prosperity. Scholars have dubbed this phenomenon, which depends on a curious matrix of economic and sociological responses to a sudden influx of petroleum money, the "paradox of plenty" or the "resource curse." Inevitably, little of the oil wealth ever makes its way to those who need it most.

One of the great scandals of the African oil boom, for example, is that it has produced far more jobs in the United States and Europe than it ever will in Africa. Only about 5 percent of the billions and billions invested in African petroleum projects every year is spent *in Africa*. Oil exploration is by its nature capital-intensive rather than labor-intensive, meaning that most investment goes to developing and operating expensive and sophisticated hardware, such as the multimillion-dollar floating production, storage, and offloading vessels (FPSOs) that have popped up along the African coastline. What little labor is needed is generally of the skilled variety, and international oil companies have scant incentive to train an indigenous workforce when flying in expatriate engineers and technicians is cheaper and simpler. Offshore oil exploration is, in the parlance of economists, perhaps the ultimate "enclave industry."

As more African economies become dependent on their oil revenues, the stakes for finding ways to beat the resource curse have never been higher. Oil and gas are already Africa's largest export category, three and a half times greater than all others put together. Extractive industries (i.e., oil, gas, and mining) accounted for more than 50 percent of African exports and 65 percent of foreign direct investment in Africa in the 1990s. The American charity Catholic Relief

Services "conservatively" estimates that $200 billion in oil revenue will flow into the coffers of African governments over the next decade. All this, it is argued, makes it more important than ever for African governments to "get it right" and ensure that oil is allowed to be a blessing rather than a curse for their long-suffering people.

Others, of course, prefer to see Africa's oil boom as simply a centuries-old story of foreign exploitation and the subjection of Africa's people at the mercy of voracious commercial interests—a second great "Scramble for Africa," after the original carve-up of the African continent by European colonial powers in the late nineteenth century. Virtually everywhere in Africa today, Chinese, Malaysian, French, Australian, and American firms can be seen jockeying for position, trying to snap up exploration acreage in an undignified rush that seems to grow more ruthless by the day. And the chess game among oil companies is inevitably echoed in the foreign ministries of the world's great powers. France, China, and the United States are engaged in an intensifying competition for influence among the oil-producing nations of Africa. China, especially, has shown that it is prepared to plunk down large cash incentives in the form of loan guarantees in exchange for lucrative oil concessions from African countries.

So who are we to believe? The committed evangelists who tell us African oil can be a catalyst for the continent's development as well as a crucial source of Western energy security? The constructive critics who stress the importance of sound fiscal management and revenue transparency, and warn of the dangers of runaway oil bonanzas? Or the Afro-pessimists who say that experience has taught us that rapacious foreigners motivated by their interests in extractive industries will only stand in the way of Africa's development? In the long run, we all know that simply drilling for oil in a faraway country without thinking about the consequences of our inevitable involvement in its internal politics is not a recipe for stability, or even for the energy security we so badly crave. That is a lesson we have probably learned from the Middle East. But does that mean we should think of Africa's booming oil production as fundamentally a blessing disguised as a curse or a curse disguised as a blessing?

It is not a question with an easy answer, nor one that lends itself to glib polemics about "blood and oil." It is, however, a question that cries out for an answer, or at least a sensible debate. For every day that goes by without such a debate is another day that MD-11s land on tropical tarmacs, another day for FPSOs to load their precious cargoes, another day of Africans becoming frustrated by their misery and suffering as they watch their leaders line their pockets with oil money, and another missed opportunity to stop matters from falling into the wrong hands.

CHAPTER I

THE ONSHORE EFFECT

"I BET YOU DON'T dare touch the salad."

It was my first week in Africa, and I must have looked every inch the amateur because I was being teased mercilessly.

"It's fine, you know. It's not going to make you sick. Not like the salads you get in London."

I was at lunch with Adwoa Edun, a Ghanaian-born, half-British owner of a Lagos bookshop who also happens to be married to a senior politician in the Lagos State government. Between lashings of gentle mockery, she was giving me her perspective as an expatriate African who had made Nigeria an adopted homeland.

"Nigerians have a tolerance level that is beyond any I have ever seen," she said. "You know, living in Nigeria, there are so many times when I have thought to myself, okay, this is it, Adwoa. We are going to have to pack our bags now. Where are we going to go? But then, every time, the country just somehow muddles through."

Sooner or later, every expatriate conversation about Nigeria comes around to some version of this conclusion—that here is a country with an unparalleled knack for survival, an almost inspired ability to lurch from crisis to crisis, even to the point of what to outside eyes resembles anarchy, before retreating from the brink and sliding back into a low-intensity seethe.

Most such conversations then turn to the subject of oil, and the volatile politics of the Niger Delta. Ours was no exception. Adwoa had no special expertise on the matter, but I had declared my intention to visit the Delta, so she agreed to give me a little friendly advice. She took my pen and notepad, and drew three large dots about an inch apart, which she labeled "Benin City," "Sapele," and "Warri." Bisecting Sapele, she drew a pair of faint wavy lines. On the Benin side of the wavy lines, she wrote PEACE. On the Warri side: TROUBLE.

"Living in Lagos," she said, "this is all we ever really know about the Delta. That if you head southeast, there is only so far you can go before you start to run into trouble."

"Trouble," in my limited experience, is a word people use when they are trying not to say "war." Growing up, I heard the conflict in Northern Ireland described by successive British governments as "the Troubles," before it was finally put to a peace process. It's one of those words, like "inclement" or "unhygienic," all middle-class and squeamish, that masks the true extent of the lurking horror. It's the kind of word that stops a conversation before it starts to get awkward; that signals, with a flick of the eyebrow and the tapping of a pen, that no more questions will be entertained today, thank you very much. "There's been some trouble," to the foreign journalist in Africa, generally means the shit has hit the fan.

By anyone's definition, the Niger Delta today is a place of troubles. Gangs of teenagers cruise the creeks and swamps in speedboats, bristling with automatic weapons. Oil is sucked out of pipelines under cover of night and sold on the black market to raise money for rival warlords. Foreign oil workers are routinely kidnapped and held for ransom. Flow stations and other oil installations are attacked and vandalized, and a general climate of impunity infects the most mundane of interactions.

Trying to untangle—much less convey—the complexities and contours of the troubles in the Niger Delta could easily become the work of a lifetime; but, as with most human conflict, its causes can be boiled down to money, land, and ethnic rivalry. The Niger Delta is made up of nine states, 185 local government areas, and a population of 27 million. It has forty ethnic groups speaking 250 dialects spread across 5,000 to 6,000 communities and covers an area of 27,000 square miles. This makes for one of the highest population densities in the world, with annual population growth estimated at 3 percent. About 1,500 of those communities play host to oil-company operations of one kind or another. Thousands of miles of pipelines crisscross the mangrove creeks of the Delta, broken up by occasional gas flares that send roaring orange flames into the already hot, humid air. Modern air-conditioned facilities sit cheek by jowl with primitive fishing villages made of mud and straw, surrounded with razor wire and armed guards trained to be on the lookout for local troublemakers. It is—and always has been—a recipe for disaster.

The problem, in a nutshell, is that for fifty years, foreign oil companies have conducted some of the world's most sophisticated exploration and production operations, using millions of dollars' worth of imported ultramodern equipment, against a backdrop of Stone Age squalor. They have extracted hundreds of millions of barrels of oil, which has been sold on the international market for hundreds of billions of dollars, but the people of the Niger Delta have seen virtually none of the benefits. While successive military regimes have used oil proceeds to buy mansions in Mayfair or build castles in the sand in the faraway capital of Abuja, many in the Delta live as their ancestors would have done hundreds, even thousands of years ago—in hand-built huts of mud and straw. And though the Delta produces 100 percent of the nation's oil and gas, its people survive with no electricity or clean running water. Education is patchy, with one secondary school for every 14,000 people. There are few public services available in the Delta, and those that do exist are difficult to reach because there are no roads. Seeing a doctor can mean traveling for hours by boat through the creeks.

Occasionally, oil has been spilled into those creeks,* and fishing communities disrupted, dislocated, or plunged into violent conflict

---

*In October 2006, the World Wildlife Fund reported that up to 1.5 million tons of oil has been spilled in the Delta over the past fifty years—the equivalent of one *Exxon Valdez* disaster every twelve months.

with one another over compensation payments. When the people of the Delta have tried to protest, they have been bought off, set against one another, or shot at. The rampant criminality, lawlessness, and youth unrest that have plagued the Delta as a result are perhaps technically "troubles" rather than active warfare, of the kind that makes the evening news and furrows brows at dinner parties. But to those who eke out a meager living in the sweltering, isolated fishing villages in the swamps and estuaries of the Delta, caught between the security forces hired by international oil companies to guard their multimillion-dollar networks of pipelines and flow stations, the roving bands of angry ethnic militias determined to disrupt their operations, and the soldiers and special police units of the Nigerian state—all sides armed to the teeth—the distinction is largely academic. On a good day, they will push off into the morning mist in their hollowed-out wooden pirogues and return in the evening with a few sickly-looking croaker and catfish that they will dry in the sun for another day.

On a bad day, they might not come back at all.

Even the most conservative estimates of the death toll—perhaps a thousand people every year—nudge it into the category of "high-intensity conflict," alongside such better-known hot spots as Chechnya and Colombia. In March 2005 the Niger Delta's seemingly intractable problems had become so severe that it prompted the U.S. National Intelligence Council to identify the "outright collapse of Nigeria" as one of the most significant "downside risks" threatening the stability of all of sub-Saharan Africa in coming years.

◄◄—►►

The largest denomination of banknote in Nigeria is the 1,000-naira note, worth (at press time) a little under $4. On its front is a portrait of Nnamdi Azikiwe, the Igbo nationalist leader who helped broker Nigeria's independence from Britain in 1960. On the reverse is a picture of an oil derrick. Both images are a reminder of a time when Nigeria seemed to stand on top of the world—sure of itself, confident about its future, and firmly in control of its destiny. It was the age of African liberation, when it felt as if a new country was being born every few weeks, its heroes brimming with promises about life without the yoke of colonialism. Nigeria, with an enormous population sitting on some of the world's largest hydrocarbon deposits, seemed poised to become an African superpower.

Along the way, though, something went terribly wrong. The country's overall economy has shrunk and the standard of living of its 130 million people declined steadily since independence, to the point where the World Bank now ranks Nigeria as one of the world's twenty poorest countries. Today, in a country that pumps more than 2 million barrels of oil a day and has the distinction of being the world's seventh-largest oil producer, 57 percent of the people live on less than $1 a day. And that percentage rises to 70 percent in the Delta. Even gasoline, which should be cheap and plentiful, is instead almost entirely imported from abroad at great cost, thanks to the near-total collapse of Nigeria's refineries.

Endless cycles of debt and crippling hyperinflation have turned everyday life for Nigeria's citizens into a painful battle for survival. There was a time when 500 naira would have seemed an enormous amount of money to almost any Nigerian, trading for around $2,000 on the currency markets. Today that 500-naira note, with the pictures of Azikiwe and the oil derrick, rarely covers a cab fare. The grubby, disintegrating brown banknotes are stuffed into glove compartments across the country like the piles of loose change that they are.

Any discussion of African oil has to start in Nigeria—not only because it is far and away Africa's largest oil producer, but also because it has Africa's longest and most exhaustive experience with international oil exploration. Oil is woven into the fabric of the nation's forty-six-year history in much the way that the faces of its liberation heroes are woven into its banknotes. However, rather than becoming a shining beacon to its less-experienced neighbors, a kind of living, breathing African oil university writing the textbook as it went along,

Nigeria instead became a case study in the sort of chaos and destruction that an oil boom can wreak on an otherwise promising nation. Across the continent, the word "Nigeria" has become shorthand for what everyone wants to avoid when they drill for oil in Africa—a synonym for "troubles."

How did things get so bad? How did a country that at one point provided more oil to the United States than did Saudi Arabia allow that very oil to become an accessory to its slow and steady unraveling? How did a lush, swampy river delta home to quiet tribes of fishermen in wooden boats become the scene of a conflict so violent and unpredictable that neither multinational petroleum companies nor one of Africa's most powerful armies seems capable of contending with it? How—to borrow a phrase from one of Nigeria's most famous novelists, Chinua Achebe—did "things fall apart"?

In 1900, the *Economist* criticized the British government's decision to annex the Niger Delta, calling it "a malarious swamp, which will cost several times the actual worth of its product." The "product" the newspaper referred to was not crude oil but palm oil, which was then prized both as a machine lubricant for the factories of the Industrial Revolution and as a base material for the manufacture of soap, candles, and margarine. The *Economist* could never have predicted that, a century later, the malarial swamp would have generated more than $300 billion in revenue from a very different kind of oil.

But the sentiment proved to be prescient.

In 1865 the British government, acting under pressure from Liverpool palm-oil traders who feared having to share their lucrative African trade with French and German rivals, had declared the Delta a British protectorate, eventually known as the Protectorate of Southern Nigeria. The use of the term "protectorate" was not accidental. The various kingdoms of the Delta had "agreed"—often in rather dubious circumstances that involved bits of paper they scarcely understood—to allow the Royal Navy to be responsible for their collective security; in effect, ensuring that no other European country could do business with them. Strictly speaking, they were unequal trading partners under foreign military protection, never "colonies."

In the decades that followed, the vast Muslim caliphates north of the Benue River—which enjoyed strong cultural, religious, and mercantile ties to North Africa and the Arabian peninsula and historically had little to do with the Delta kingdoms to their south—were also "subdued" by the British, who, in recognition of their hierarchical and more scientifically advanced civilization, granted them an unusual amount of autonomy under the umbrella of the Protectorate of Northern Nigeria. In 1914 the southern and northern Protectorates were brought together with the Lagos Colony to the west (which had been set up primarily as a forestry concession), and the three were amalgamated into what a British governor's wife had once suggested could be called "Nigeria."

The Protectorate of Southern Nigeria had been dominated by the Igbo (or Ibo), a 15 million–strong tribe of yam and cassava farmers who maintained uneasy relations with the Delta's dozens of minority tribes (at least one of which, the Ijaw, had sold them into slavery for the better part of a century). In the north, the mostly Muslim Fulani Hausa held sway, while Lagos and the Southwest belonged mostly to the Yoruba. After independence, the prevailing assumption was that, in the interest of national stability, the three regions would find ways to share major government positions between them. This informal arrangement soon became known as the "federal character principle."

In fact, Nigerian politics since 1960 could be described as an uneasy, loveless ménage à trois between the three majority tribes, each of which believes the other two are working in cahoots against it, while the country's 200-plus minority tribes feel left on the sidelines to fend for their survival as the big three divvy up the spoils of the country. To make matters more complicated, each majority tribe regularly forges political alliances with minority tribes in other parts of the country, playing on their frustrations to strengthen its position and appear "patriotic" and pan-Nigerian, while undermining its majority-tribe rivals in their own backyards. In a perverse way, though, this system of ethnic horse-trading has kept Nigeria together over the years. In the 1960s Nigerians learned the hard way that ethnic separatism, although seductive, could lead only to death and destruction in a state as balkanized as the newly independent republic.

The first rumblings of trouble in the Niger Delta began in 1966, when the Ijaw, one of Nigeria's largest minority tribes, realized they were sitting on a gold mine. In 1956, in the Ijaw village of Oloibiri, Shell had discovered oil, and Nigeria's output soon soared to more than 400,000 barrels per day, much of it extracted from the swamps and creeks of Ijaw country. Already resentful of Igbo dominance in the southeast, but keenly aware that the Northerner-dominated government in Lagos (then the national capital) was unlikely to use oil revenues for the benefit of the Ijaw, Isaac Boro and Nottingham Dick, two idealistic young radicals, founded the Niger Delta Volunteer Service in February 1966 and declared the Delta an independent Ijaw republic. The provisional "government" of Ijawland pronounced all oil contracts null and void, ordered oil companies to negotiate directly with the new republic, and told all non-Ijaws to register with the NDVS within twenty-four hours. The NDVS managed to capture Yenagoa, the area's largest city, before the Nigerian army moved in, using pontoon boats on loan from Shell. The fledgling republic was quickly quashed, but Isaac Boro went down in Ijaw history as a hero who had fought the Nigerian state on behalf of his people. From the perspective of the federal government, a dangerous precedent had been set.

The three years that followed were the most tumultuous and painful in Nigerian history. The Biafran war of 1967–70, touched off when the Igbo declared an independent Republic of Biafra in the southeast of Nigeria, was the world's first televised African tragedy and the beginning of the end for the euphoric Afro-optimism of the 1960s. For months, the world watched footage of starving children as the Igbo claimed (not without a little hyperbole) that the Nigerian state was perpetrating a "genocide" against them. According to some estimates, two million people died during the war, the majority from disease and hunger. There are many reasons why the Biafran Republic was unable to break away from the Nigerian federation, but it didn't help that the Igbo were never able to rely on the support of important southeastern minorities, such as the Ijaw, who knew all too well that they would suffer a far worse fate in an independent, Igbo-dominated Biafra.

The Biafra episode effectively scuppered for decades any hotheaded talk of secession and ethnic separatism. Though frustration with the federal government continued to mount among the Niger Delta minorities in the 1970s and 1980s, few protests ever reached the point of violent standoff between activists and Nigerian troops. Disaffected local youths who did decide to cause trouble often failed to elicit support from their own communities, who preferred to keep their heads down and focus on survival. No one in Nigeria had the stomach for another Biafra.

These were also the years in which international oil companies—uninterested in longterm cohabitation with the local communities, or unsure of how to achieve it—made it their unofficial practice to pay off village chiefs to ensure that local youths did not disrupt their operations. This approach seemed to work for a while, but ultimately succeeded only in creating violent disputes between neighboring villages vying for oil-company handouts, not to mention ugly contests for the suddenly lucrative title of village chief. Centuries-old traditions turned into naked money grabs, as traditional rulers simply pocketed oil-company handouts and proved unable to contain angry youth. Exasperated, the oil companies turned to the youths themselves, offering them "ghost jobs"—which required nothing more from them than a promise not to attack oil installations. They were literally paid to stay at home.

Little by little, foreign oil companies found themselves trapped by the twisted but irrefutable logic of it all, internalizing its assumptions and learning to accept—sometimes reluctantly and sometimes willingly—a degree of moral compromise as part of the price of doing business in Nigeria. The Nigerian government, for its part, keenly aware of the importance of oil revenue to its own survival, maintained an iron grip on power and an attitude to law and order that made the oil companies look like gormless pantywaists. It was, in effect, a carrot-and-stick approach that kept the Delta quiet in these years. On the one hand, the somber specter of a bloody Biafra-style war of attrition lurked as a powerful deterrent against any large-scale organized uprising; while, on the other hand, rampant bribery ensured that the people of the Delta understood that capitulation carried its own rewards. Trapped between the Scylla of moral bankruptcy and the Charybdis of ethnic cleansing, only the most stubborn of ideologues talked of the rights of their people. The Delta lapsed into an uneasy equilibrium.

By the beginning of the 1990s, however, the situation was getting out of hand again. In the 1990 "massacre" at Umuechem, scores of people from the Etche tribe were allegedly killed by Nigeria's infamous Mopol (Mobile Police) forces—nicknamed "Kill and Go" for their lack of interest in kid-glove policing. On October 29, Shell managers, who had heard word of an "impending attack" on their operations near Umuechem, asked the Rivers State Police Commissioner to send antiriot police to protect their facilities. The "attack" turned out to be a peaceful protest outside the Shell installation, but the requested Mopol unit opened fire on villagers, most of whom scattered into the surrounding bush. For good measure, the police returned just before dawn the next morning and slaughtered those villagers they found returning from the bush. According to Amnesty International, 495 houses were damaged or set on fire and 80 people killed. The subsequent judicial inquiry, in a display of independence rarely seen during the military regime of the time, ruled the police had shown "a reckless disregard for lives and property."

It was a pattern that would be repeated throughout the Delta in the 1990s. Years of compromise and stagnation would give way to spasms of anger. Shell, or one of the other major oil companies operating in the area, would find itself the target of demonstration and would seek the protection of the Nigerian authorities. And such protection would arrive in the form of overenthusiastic, underpaid young soldiers—some of them from tribes with a history of animosity toward the offending parties—commanded by officers who had received a wink and nudge from higher-ups suggesting that the communities should be made to pay for their audacity.

Again and again, in village after village, the same sorry scene would be acted out, as military junta met defenseless citizen in carnivalesque sprees of violence and retribution that would invariably leave bodies of sons and grandmothers writhing in the mud. Every time, the guns would fall silent, and the fishermen would go back to the creeks. And every time, international human-rights organizations would issue obligatory condemnations. Occasionally, they would write more detailed reports, filled with chilling details about the level of brutality involved, but still few outside Nigeria took notice of the degenerating situation in the Delta.

Then along came Ken.

In a region of small, overlooked minorities, the Ogoni were one of the smallest and most overlooked. The entirety of what the more militant Ogoni would refer to as "Ogoniland" is only 400 square miles in area, with at most 500,000 people—a negligible presence in a country of 130 million. But by the early 1990s, Shell's ninety-six wells had pumped over 600 million barrels of oil out from under Ogoniland, and that oil had sold for billions of dollars on the global market. As early as 1970, six Ogoni chiefs had written to the governor of Rivers State, asking for a greater share of oil revenue and a redressing of environmental damages caused by oil exploration. Their letter had gone unanswered.

In late 1992, a group calling itself the Movement for the Survival of the Ogoni People (MOSOP) issued a thirty-day ultimatum to Shell, demanding that the company either pay back-rents and damage compensation to the communities affected by its operations or prepare to leave Ogoniland for good. MOSOP was led by the charismatic Ken Saro-Wiwa, a journalist, novelist, and soap-opera scriptwriter with a knack for attracting publicity. Earlier that year, Saro-Wiwa had traveled to Europe, where he had made contact with prominent social and environmental activists such as Body Shop founder and CEO Anita Roddick. When the MOSOP ultimatum expired in January 1993, 300,000 Ogoni staged a peaceful demonstration that went off largely without incident. Shell's managers in London and The Hague realized they had a problem on their hands.

Three months later, in April, employees of the American pipeline company Willbros, under contract to Shell, were confronted outside the village of Biara by a group of Ogoni farmers who demanded that they desist from their work. Nigerian army troops quickly arrived on the scene and began shooting at the activists, killing one and injuring eleven. Shell subsequently announced that it had been forced to suspend its activities in the area because of public hostility—an admission that sent a ripple of dramatic tension throughout Nigeria and raised the specter of "another Biafra." The parliament in Abuja branded MOSOP a secessionist and treasonous movement, and banned it.

During the year that followed, increasingly violent clashes— some of them between Ogoni and neighboring tribes, secretly provoked by the Nigerian army—left hundreds of Ogoni villagers dead. All the while, Shell itched to return to work in the fields it had abandoned after the Willbros episode. Finally, in May 1994, events came to a head when a meeting of Ogoni leaders was disrupted by a mob that appeared out of nowhere and murdered four of the tribal chiefs. Ken Saro-Wiwa, who most observers now agree was nowhere near the meeting, was arrested and later charged with the murders, along with eight other MOSOP activists. In a series of secret meetings held at the home of Saro-Wiwa's brother, Owens Wiwa, in July 1995, Shell's then-chief executive for Nigeria, Brian Anderson, offered to intercede with the authorities on the nine men's behalf, on the condition that MOSOP end its campaign and put out a press release absolving

the company of responsibility for environmental damage in Ogoniland. The Wiwas flatly refused to hand Shell the PR victory it wanted
so badly, and on November 10, 1995, following a trial that observers
condemned as a farce, the Ogoni Nine were hanged from a gallows in
Port Harcourt. The news was greeted with shock and disbelief by the
international coalition of activists that had flocked to the Ogoni cause
over the previous two years. British Prime Minister John Major described the execution as a "judicial murder," and Nigeria was immediately suspended from the Commonwealth.

By the late 1990s, things in the Delta had spiraled steadily out of
control. The peaceful, disciplined protest movement put together by
Saro-Wiwa and MOSOP had been replaced by a far more sponta-
neous and confrontational style of activism that viewed criminality
and vandalism as justifiable weapons in a guerrilla war against the
Nigerian state. Gangs of disaffected youths saw no shame in occupy-
ing flow stations, sabotaging pipelines, and kidnapping—or even
killing—foreign oil workers. The most desperate took to vandalizing
pipelines in the hopes that the subsequent oil spill would result in a fat
compensation check for their villages. And, for the first time, there
emerged organized syndicates whose stock-in-trade was tapping into
the production line and stealing crude oil to sell on the black market—a practice that has since come to be known as "illegal bunkering." By 2003 an estimated 200,000 barrels of oil was disappearing
every day in Nigeria, causing a loss to the national treasury of some
$100 million a week.

Perhaps the most worrying development, however, was that, increasingly, disenfranchised communities were spending less time in
confrontations with oil-company security personnel or Nigerian soldiers, and more time clashing with one another. At issue, generally,
was the right to be recognized by international oil companies as an
"oil-producing community." Over the years, villagers had seen that
such a designation carried with it an intoxicating raft of privileges.
Every time an oil company wanted to drill in a new patch of the Delta,

international law and its own corporate guidelines would require it to undertake a social and environmental impact assessment to determine potential disruption to the local community. The company would be expected to meet with community leaders and listen to grievances after operations had begun. There would have to be an effort to provide jobs to local youth. And if there was ever an oil spill, the company would have to pay compensation.

In a region that national politicians had neglected and exploited for decades, locals understood all too well that the white men with the drills were their last, best hope for the development they had expected the oil wealth to bring. With their whizbang technology and can-do corporate spirit, the oil companies came to be seen as a surrogate for the state. It was a role they neither relished nor were particularly qualified to perform. In the marshy tidal waters of the Delta, where human settlement patterns have tended to follow the catches of fishermen, village boundaries have never been clearly defined by local authorities. Inevitably, oil companies' attempts to identify and deal with community representatives fell foul of territorial disputes. Such disputes are nothing new to the Delta, but traditionally they have been over fishing rights. When the stakes became thousands of dollars in oil-company handouts, the results were predictable. In March 1997, when the headquarters of a local government area in Warri was moved from an Ijaw town to an Itsekiri town, bloody riots broke out. In 2003 the dispute reignited, and the ensuing violence left nearly one thousand people dead. Chevron was forced to shut down its operations at the nearby Escravos terminal, with the result that 800,000 barrels of oil a day—one-third of Nigeria's output—was taken off the international markets for several months.

The Ijaw have always been the largest ethnic group directly affected by oil exploration in the Delta, so it is no surprise that, in the years after Ken Saro-Wiwa was hanged, the mantle of resistance passed back to the Ijaw. Inspired by the outpouring of international sympathy that a tiny ethnic group like the Ogoni had been able to muster, the much more populous and much more radicalized Ijaw began to organize. On December 11, 1998, Ijaw leaders gathered at Kaiama and, in an echo of the Ogoni ultimatum of six years earlier, declared that all oil companies were to leave by December 30, "pending the resolution of the issue of resource ownership and control in the Ijaw area of the Niger Delta." When December 30 came around, security forces fired on Ijaw youth seeking to implement the terms of the Kaiama Declaration in the Bayelsa State capital, Yenagoa. A one-week state of emergency was declared as Nigerian troops and Ijaw youth fought running battles throughout Bayelsa.

The tragedy of the Niger Delta story is that it could be told through the eyes of any one of the many Delta minorities affected by oil production. Urhobo, Ijaw, Etche, Itsekiri, Ogoni, Edo, Efik—all have some version of the sorry tale to tell. When I visited Nigeria in January 2005, however, the Ijaw community of Kula was in the news. A few weeks earlier, angry that Shell's and Chevron's promises of development projects had not been fulfilled, thousands of Kula villagers had occupied the companies' flow stations in the area, shutting in 120,000 barrels of oil a day. The protesters had refused to leave until a new Memorandum of Understanding (MoU) was signed, with clear guarantees of compensation and infrastructure projects for the community.

Over the years, MoUs have become standard operating procedure for international oil companies and local communities, who know that dealing with each other directly is infinitely preferable to leaving things to the Nigerian government. Unfortunately, MoUs are, by their nature, informal documents, outlining generally agreed-upon principles, and rarely amount to much more than a handful of promises—such as financing the construction of boreholes or clinics—made by an oil company in exchange for a peaceful operating environment. Routinely, when communities feel promises are not being met, they take over flow stations or otherwise sabotage operations in an attempt to draw attention to the problem.

After several weeks of shut-in production in Kula in December 2004, the dispute had been resolved thanks to some heavy-handed intervention by the Rivers State governor (and probably some money thrown at the village chief), but tensions were still running high. Kula elders were threatening to make life hell for Shell and Chevron, and no one doubted the potential for violence.

And so it only made sense to pay a visit to Kula.

-◄-◄-►►-

Nearly everyone who visits the Delta begins by flying into Port Harcourt, the unofficial oil capital of Nigeria. There are daily flights from London, Paris, and Houston, as well as dozens a day from Lagos and Abuja. Outside the airport, fleets of gleaming, air-conditioned SUVs, their engines purring in the stifling heat, sit waiting to receive sweating politicians and oil-industry personnel as they work their way out of the screaming chaos of the arrivals hall. The flight from Lagos takes about an hour and I had been told by everyone that it was the only reasonable way to make the trip—that, despite Nigeria's appalling civil-aviation safety record,* trying to go by land would be a mark of total insanity.

But I was determined not to fly. I kept thinking of the wavy lines Adwoa had drawn on my notepad between Benin City and Warri, and it seemed a shame to traverse them from thousands of feet in the air. I wanted to know what "trouble" looked like on the ground. And so, at seven o'clock one sweltering morning, along with eight other passengers and mountains of luggage, I piled into the back of a decrepit early-1980s Peugeot 504 operated by the Edo State government.

---

*Within just a few weeks in late 2005, three embarrassing accidents claimed the lives of more than 200 people in Nigera. One plane burned for hours on the runway at Port Harcourt because the airport had no functioning fire engines, killing over 100 people. Another flight crashed and killed 117, and it took authorities fifteen hours to find the crash site. In a third incident, an Air France jet plowed into several head of cattle that had strayed onto the runway in Port Harcourt. The offending livestock were quickly scooped up by airport employees who made an impromptu barbecue from the carcasses.

It took about two hours for the quorum of passengers to turn up, and longer for them to buy their tickets, check out the roadworthiness of the car, and make official their commitment to the journey, during which time we early birds sat sweating into the mangled wooden benches and loose springs left from the vehicle's once-plush vinyl seats. Every ten or twenty seconds, an arm would reach through an open window and dangle tampons or sausage rolls or bicycle chains in front of my face, holding them there until I could firmly refuse the offer.

With the car filled, it looked like we were ready to go—but not before obtaining a sort of insurance. A heavily perspiring preacher appeared at the open door with an open Bible pressed against his lips, the edges of its crinkly pages turned a rich dark brown from years of spittle. At the top of his lungs, and at the breakneck speed of a horse-race announcer, he prayed for all our souls: *"Lord in the name of the holy savior Jesus Christ we pray for these passengers we pray for you to deliver them safely to their destinations we pray for their luggage and ask you to protect them and let them survive the journey we pray for the driver and ask that you watch over him and the other drivers on the road we pray for the car he is driving and pray for its engine and its suspension and gears and axles and ask you to watch over it we pray for the condition of the road."* And on and on until every potential hazard of the journey ahead had been accounted for and cleared with the Almighty. After we had muttered a dutiful "amen," the preacher slammed the door and we were off.

Traveling by road in Africa is never a relaxing experience, but the journey into the Delta from Lagos literally bleeds with misery. Waves of human suffering lap against the edges of the road, periodically splattered back into place like stagnant puddles under the wheels of passing cars. Dystopian landscapes flashed by the windows like the pages of a demented Dr. Seuss book, or an Old Testament catalog of ordeals. Heaps of burning garbage, some as tall as buildings, sent flames and smoke and swirling ash into the already stifling air of the car. Lepers wrapped in bandages—ostracized from their villages and unable to obtain work—rushed up to passing cars, waving crude handmade flags to warn of potholes, in the hope that motorists would fling loose change at them before they got too close. As for the road, it was little more than an endless stretch of stray boulders, man-sized craters, and rivers of open sewage, punctuated every few hundred yards by semiofficial "roadblocks"—generally just two rubber tires, a pile of sticks, and a pair of policemen wielding mangrove branches and shaking down motorists for loose change.

Swerving violently to avoid potholes and oncoming trucks, our driver popped a gospel tape into the cassette deck, and began humming along quietly. The fashionably dressed woman in her early twenties sitting next to me looked up from her book of hard-core erotic fiction and led the others in a gentle rendition of the hymn "I Know Jesus Is My Savior." About an hour into the revival session, when our little improvised traveling choir had transformed into a boisterous, clapping octet, she drifted off to sleep, her spiky coif nuzzled against my face and her book open to a detailed narrative of cunnilingus.

By the time I was dropped off in front of my hotel in Port Harcourt, both legs were asleep, and I was sporting several bruises from the elbows, door handles, and luggage buckles that had been pressed into my flesh as the car dropped in and out of craters most Western SUV-drivers wouldn't dare tackle. A journey of 350 miles—approximately the distance from New York to Boston—had taken two full days. As the car pulled away in a clatter of loose bodywork and a pillow of black exhaust, I caught sight of the Edo Lines slogan pasted onto its windshield: "Jesus, lead us."

The Son of God, I soon discovered, was in high demand in Port Harcourt. Nigeria, like most of Africa, is deeply religious, but in the Delta, the charismatic evangelical brand of Pentacostal Christianity, much of it imported from the United States, has been particularly successful at finding adherents. In some parts of Port Harcourt, every other building seems to be a church—or rather, a "ministry"—and every few hours you hear clapping hands and peals of boisterous worship.

And it's not hard to see why messages of miracle and redemption have found a ready audience. Here, where the overwhelming majority of people have nothing, epic helpings of a highly visible wealth have been bestowed on poor neighbors and friends seemingly overnight. In a more meritocratic corner of the world, the poor might believe they're poor because they haven't worked hard enough. In the Delta, though, where there is little connection between hard work and fabulous wealth, it is much easier for the poor to believe they're poor because they haven't *prayed* hard enough. Here, TV sets seem perpetually tuned to American game shows where somebody's always winning a new car, or to evangelical broadcasts where grinning American women with big hair and plastic cheeks are praying for our souls.

If Abuja, Nigeria's purpose-built federal capital district, is its Washington, and Lagos, its manic and multicultural commercial hub, its New York, then Port Harcourt must be Nigeria's Houston. A tangle of fading expressways and overpasses that sprang up during the oil boom of the 1970s, the town is surrounded for miles by giant natural-gas flares and billboards flogging holy salvation. It's a raw and rough-hewn town, a place of hustlers and hawkers and white men in SUVs, where the neighborhoods bear such exotic names as "Trans-Amadi Layout," "GRA Phase II," and "D/Line." And, in a country that has become famous around the world as a nerve center for hucksterism and scams both felonious and petty, Port Harcourt's finest confidence men have come up with what must be the mother of all frauds: "selling" houses that don't belong to them. All over town, nervous homeowners have resorted to painting THIS HOUSE IS NOT FOR SALE on their exterior walls, as a warning to hapless househunters who might otherwise find themselves being given a guided tour of the premises while the owners are away.

Yet despite the ostentatious wealth of a lucky few, the city is unmistakably poor—desperately, grindingly poor. The hundreds of thousands of people who flocked to Port Harcourt during the oil boom—turning it from a city of 200,000 to one of 1.1 million within a few years—have been joined on the streets by thousands more young people from all over the Delta who continue to treat "Potako" as a default destination when all else has failed. The Delta's youngest and brightest run alongside moving cars on the expressway, in the suffocating tropical heat, in the hopes that someone will roll down a window and buy a pen or a phone card or some AA batteries off them, before one of the legions of legless panhandlers has had a chance to bang on the car door. It's not a place to spend your honeymoon.

Kula, like much of Ijaw country, is not reachable by road. A people whose destiny has been tied to tidal fishing for centuries, the Ijaw live on top of steamy, spongelike mangrove swamps that rarely climb to more than three or four feet above sea level. They live precariously at the best of times, in mud and straw huts that seem to hover over the water. A patch of rough weather at sea can wash away a village in a matter of hours. Picture an African Louisiana without a levee.

And, like much of Ijaw country, Kula is nowadays considered unsafe territory for a white man traveling alone, thanks to increased militant activity and widespread anger at foreign oil companies. Even if you managed to negotiate a reasonable rate for boat hire, the warning goes, you would have a hard time convincing village youth that you were not an oil-company worker and should not be taken hostage. Only a few weeks before my visit, a foreign journalist using an inexperienced guide had been kidnapped and held in the creeks for several hours. I was painfully aware, therefore, that I would need a guide I could count on. Someone with some real star power, who knew the creeks like the back of his hand. Someone who spoke the local tongue, who knew how to sweet-talk an armed militia, and who wouldn't lose his cool in a sticky situation. It took a few phone calls to pin him down to a meeting, but with a little persistence, I was soon face to face with just such a man: the one and only Felix Tuodolo.

In the late 1990s, Tuodolo had founded the Ijaw Youth Council, a vanguard of radicalized youth that he had tried to channel into a constructive, coordinated advocacy network along the lines of MOSOP. It had made him one of the most respected and credible voices for change among the Ijaw. Now in his mid-twenties and studying for a Ph.D. in conflict studies at Liverpool University, Tuodolo was back in Port Harcourt on winter break, and asked me to meet him in the lobby of the Presidential Hotel.

"The lobby of the Presidential Hotel" is itself a timeless African cliché—the kind of place immortalized in Graham Greene novels and James Bond films, where figures with many zeros are whispered over glasses of Chivas Regal, and diplomats' wives come to get their hair done. Sit in one of its leather armchairs for more than a few minutes and you will be treated to a nickelodeon of passing kingmakers and rogues, from cabinet ministers to wildcatters, UN technocrats to foreign correspondents. It's the kind of place journalists can come to and take care of all their interviews in the course of an afternoon, over a steady succession of scotch-and-sodas and delicate insinuations. The kind of place some people never leave.

As I sat staring blankly at the farrago of Louis Vuitton cases and quiet pinstripe handshakes silhouetted against the lobby's automatic doors—on the other side of which a conveyor belt of taxis and chauffeured cars unloaded their self-important cargo at what seemed like perfectly timed intervals of forty-five seconds—I began to wonder if Felix was ever going to turn up. Then, from behind me, I heard an ironic voice declare majestically, "The president emeritus of the Ijaw Youth Council!" to peals of laughter and the enthusiastic slap-slap of a round of athletic handshakes. I turned to find a slim young man in a tight white T-shirt and wire-frame spectacles grinning modestly and trying to swat away his fame.

When the adulation had subsided, Felix and I withdrew to the bar and tried to have a conversation, but we were constantly interrupted by calls to his cell phone from young activists asking him to take sides in various disagreements. Then, as I started to explain my work, a group of young men approached and launched into what seemed like a plaintive-sounding speech in Ijaw that ended with Felix reluctantly giving them a little money. More phone calls and more handshakes followed, and an exasperated Tuodolo finally asked me just to meet him the next morning for a trip into Kula. He would, he assured me, make all the necessary arrangements.

◄◄·►►

The road from Port Harcourt leads south, inevitably, toward the sprawling mouth of the Niger Delta. And as it insinuates its way through thick forests of coconut palms, it passes the usual heaps of acrid garbage smoldering in all their tear-jerking glory. But here, going south from Port Harcourt, the road also takes on several forms of punctuation unique to the Delta.

First there are the fences and barriers of the international oil companies, each painted in the companies' signature colors and accompanied by rusted but still-menacing signs that warn against unauthorized entry. Then there are the small wooden shacks staffed by young men selling glass bottles of black-market fuel, under the noses of the oil companies. And everywhere, everywhere, are the election posters for Rivers State governor Peter Odili, considered by many one of the world's most corrupt politicians. Sporting a broad-brimmed white trilby hat and walking stick, Odili beams out over the misery like a grinning T. J. Eckleburg, framed by the caption "The Portrait of a Performer."

After about an hour, the road peters out in a place called Abonema, at which point only watercraft can continue the journey. Felix had me wait in the car while he negotiated the boat hire. "If they see a white man," he said, "they will double the price, and you won't get them to budge."

The "speedboat" we ended up with was a simple fiberglass hull with a sputtering outboard motor strapped to its stern, but it hugged every curve as if on a racetrack, as we whizzed through channels of water at a good thirty to forty knots, slowing down only to avoid capsizing the delicate wooden pirogues paddled by fishermen, or when Felix and the driver disagreed about the best route to take. Listening to these detailed arguments over exactly which clump of mangrove trees was which, I was faintly embarrassed that the only visual aide-mémoire I found at all meaningful were the occasional pipelines and flow stations and pressure valves belonging to the oil companies. My own vernacular. Without them, and without my guides, it might have taken me years to find my way back to Abonema.

Like mice in a giant watery maze, we traveled a good twenty-five miles in anything but a straight line. Every time we sped along a broad boulevardlike waterway flanked by mangrove trees and I thought things looked easy, the driver would swerve violently into what appeared to be a riverbank but turned out to be a "shortcut," and we would blast our way through a dense alleyway of water no wider than the boat itself. The motor failed every few minutes, and then we would sit waiting for the driver to repair it, overcome by—to me— a completely alien absence of noise. Even the quietest places on earth have some sort of ambient sound—a rustling wind, chirping birds, the distant hum of fluorescent lights. But here, in the deathly humid air of the Delta, when the engine failed, and the waves stopped lapping, there was total silence—like a pitch-black blindness of the ears.

After an hour and a half, I noticed a village of mud huts, skirted by a black, grimy beach that appeared to be blanketed with several inches of garbage. As our boat drew near, a constellation of small faces looked up and paused. Faint shouts of excitement could be heard, and a crowd of young men quickly gathered along the wooden landing pier, staring with apprehension at the white man approaching in the speedboat.

The driver tied up the boat, and a dozen children ran down to see if we had any luggage they could carry. A sign on the pier welcomed us to the "Kula Kingdom." Felix introduced himself and a ripple of recognition fanned across the crowd. The chief of the kingdom was away in Port Harcourt, so we were welcomed by Nye Morine, who described himself as "coordinator of Ekulama Houses," and allowed us entry into the village after a little ceremonial negotiation with Felix, but not before letting it be known in his loudest English, "We are still suffering. Shell, Chevron have done nothing. Nothing!"

We were led up the dark staircase of a roofless, disintegrated concrete building, where we sat in a ring of plastic chairs as the men gathered around to have their say. Victor Solomon, an articulate thirty-two-year-old who described himself as a "youth leader," did most of the talking, while others nodded and prodded and whispered in his ear. It was not the first time the men of Kula had entertained the press.

"We are still suffering, look how we are suffering" was their refrain, and I was going to hear it several dozen more times during the course of the conversation. I asked about the flow station they had taken over. "The government has persuaded us in different ways to open the station," they said. "Finally, they persuaded us with the military to open it. They told us, if we don't open it, they will open it for us." There was no need to explain what that meant.

A well-rehearsed list of grievances followed:

"Forty-three years, what do you see? People dying of starvation, hunger, illness. The CLOs [Community Liaison Officers sent by the oil companies] and the managers connive to steal our compensation. When there is a spill, they give us 35 naira [about 25 cents] per net."

"There is not one Kula person working for Shell. There are people in Kula with master's degrees, Ph.D.s—we even have some who have gone abroad. There is not one Kula person working for Shell. Not one."

"We are suffering," mumbled the Greek chorus. "Look how we are suffering."

"The minute you go near a facility, they come out with guns. So you don't even have a right to demand anything from them." In 1997, Victor claimed, he was shot in the leg by the Nigerian navy on orders from Shell, after he tried to protest an oil spill.

"We have no local government."

"When I was a child, you could see fish."

"See how we are suffering."

A thin, intense-looking man named Otonye Lucky Alalibo appeared abruptly from the back of the room and handed me a letter certifying his graduation from a Shell-sponsored "Youths Training Scheme" in Bakery and Confectionery. It was dated 1997, when Lucky was still a "youth," and he said he had heard nothing since. The letter was slightly soiled and creased, but had obviously been treated with great care during its eight years in the jungle—a black-and-white reminder of dreams deferred. It praised him for being, among other things, "well-behaved."

It was back to Victor: "Shell have leased this land, our grand-fathers' land, for forty-six years, and they haven't paid even ten kobo in rent," the symbolic sum being worth around one-fifteenth of an American cent.

Tubotamola Pokubo, who called himself vice president of the Kula Youth, returned to the issue of the fishing nets, pointing out, "When there is a spill, Shell pay 400 naira [about $3.00] for a bundle of nets. A new bundle costs thousands [of naira]. They pay five naira [around four cents] a yard." Referring to the takeover of the flow station, Pokubo added, "Tell Shell to take very good care of this community, or next time we are going to blow it up. We need the White [i.e., the white man] from Shell to come here for himself."

We walked around the village, a crowd of twenty young men stay-ing uncomfortably close. When we walked, they walked. When we stopped, they stopped. I felt a bit like a spoonful of honey being waved back and forth among bees. Yellow and red oil drums painted with the Shell logo lay scattered about, some being used as work surfaces or storage containers by the women. One woman cleaned fish from a tray swarming with flies, throwing the fish into a bucket of brown water. She told me she sometimes barely made 1,000 naira a day.

In a nearby shack a hollowed-out old man sat in the dark, his eyes bulging as he whittled silently. He looked frightened by the crowd. A young man next to me piped up to say it was his father, and that he was making paddles for canoes.

They took us next to a woman weaving "fish cards"—plate-sized straw paddles to which fish are attached for drying in the sun. The woman said it took her ten minutes to make each one, which she might then sell for two naira (about one and a half cents). On a good day, she said, she could make a hundred fish cards.

The men showed me a flight of stairs that went nowhere, ending after the sixth or seventh step. They belonged to the first brick house in the village, built in 1973, which was now little more than a hollow rampart of crumbling masonry. I was surprised to hear someone still lived in it; in fact, it was a prized dwelling, for a big man.

A smiling, good-natured young man named James Sunday showed me the wood and straw house next door, where he lived. A cobra had recently made its way into the house and killed his brother. "We are fighting for our rights," James said earnestly. "To get our human rights." I later learned he meant fighting literally, as he had joined the forces of the Ijaw warlord Dokubo Asari a few months earlier, when Asari had declared war on the Nigerian state and sent the price of oil up by $2 a barrel on the international market. Ijaws believe the god Egbesu protects them from the firepower of the Nigerian military. James told me, "We wear charms and amulets that make us bulletproof."

They showed me the "community toilets," which were two wooden stalls perched above a stretch of open water. Then they showed me the "sandfill" they said had been promised by Shell, which they claimed was a job half done, and could not be built on because it would wash away in the rainy season. I looked over to see Felix crying.

We began walking to the "hospital," which they said had two or three nurses but no doctor. The nearest doctor was back in Abonema. As we walked, a soft-spoken young man named Ajemina Daniel grabbed my arm and informed me that it had been decided that this was the year they would destroy the flow station if their demands were not met. I asked if they were not worried about the navy guards with their guns. Not at all, he said. "We will consult our gods."

I was shown the Kula Women's Development Center, with the words "Fashion and Desiner" painted over one door. It had been opened last year, but apparently stayed open for only two months. I was told that the money from Shell that had been supposed to come every month had stopped coming.

They showed me a drinking well, filled with thick brown opaque water. To prove the point, they lowered a dirty yellow plastic bucket, then passed it round and drank from it. They offered it to me, but I refused.

Ajemina Daniel approached again. He showed me his license to be a quartermaster, which he had obtained when he was twenty-one. He was now thirty and said he had never worked. He began to grow angry, and clearly wanted my attention. "My grandmother died three years ago, and she is still in the mortuary in Port Harcourt. I have no money to bury her. I am so angry. I wish you weren't a journalist; we would have kidnapped you and held you here. I am so angry; I am ready to sacrifice my life. I don't care if they kill me."

Felix injected a little capital into the Kula economy by buying two *piti* (a sort of tamale made of mashed corn and plantain), which he fished out of a slimy bucket. He offered me one but I declined, though they did look nice. James laughed sympathetically and said, "You know they are not well prepared." Several men joked that if I had been with Chevron or Shell, they would have insisted I eat them.

We passed on to what they told me was a bathing pond—its water was jet black—and began to walk back to the jetty. I noticed a lone generator hooked up to a stereo blasting reggae. An older man, his breath reeking of alcohol, rushed up to ask if I was with Chevron or Shell. I explained that I was a journalist, but he seemed unsatisfied with the response, and asked if I was a journalist with Chevron or Shell. James informed me that he wanted to "cause me harm" if I was.

As we prepared to leave, amid a blizzard of grinning requests for money, I noticed that this tiny village kingdom had not one but three jetties, all right next to one another. When I asked about this, it was explained to me that the one to which we had tied our boat had been built in 1982 by Shell; another had been started in 1989 by OMPADEC (Oil and Mineral Producing Areas Development Commission, the military regime's short-lived attempt at addressing the grievances of the Delta), but had never been finished; the third had been erected in 2004 by Shell, but not officially put into commission. Like the clinic without a doctor, the schoolroom without teachers, and the women's center that quickly lost its funding, these jetties were emblematic of the compulsive, reactive Band-Aid approach to the Niger Delta's problems. Every so often, a community would grow restless and noisy, so someone from the government or the oil company would throw them a bone in the form of a schoolroom or a borehole or a jetty. And then, just as quickly, the villagers would be forgotten and left to return to weaving fish cards and squabbling over cash handouts.

For years, the holding pattern worked, at least according to its own perverse logic. Conflict simmered throughout the Delta, but it followed a template that was easy to understand: suffering, frustration, protest, organized uprising, violent crackdown, Memorandum of Understanding, token development project, discreet cash handout, return to suffering. Year after year, the same tired set pieces were acted out, with only the cast of characters changing, a sort of rotating repertory of protests and pipelines and white men arriving in helicopters.

Then, in the late 1990s, people started getting creative.

◄◄-►►

In 1993 seven cases of "pipeline vandalization" were officially recorded in the Niger Delta. A couple seemed politically motivated, a couple seemed attempts to elicit cleanup contracts from the resulting spills. The reasons for the others were unclear. In 1996 the number of pipelines vandalized was 33, and in 1998 the figure had risen to 57. Still, most cases were treated as outcomes of economic disputes or simple sabotage meant to score political points. In 1999, though, a whopping 497 instances of pipeline vandalism were recorded, and in the following year there were over 600. Suddenly oil companies had to deal with a threat to their operations that was more complicated than a few kids getting carried away. The culprits were not just damaging company property, they were stealing crude oil and selling it on the black market. By 2004 Nigeria was losing as much as 200,000 barrels of crude oil a day—nearly 10 percent of its output. An influential newsletter called what was happening in the Delta "theft on an industrial scale."

The practice, known as "illegal bunkering" (in its legal form, bunkering is the act of loading fuel onto a large ship's on-deck fuel bunker), involves tapping into a pipeline, filling plastic jerry cans with crude oil, and taking the oil away in speedboats to waiting barges, which in turn sell the oil to large oceangoing tankers, which then sell it to refineries in neighboring countries such as Ivory Coast, at a considerable profit.

This is not a job for amateurs. Multinational oil companies tend not to invest in pipelines that can be sliced through with an ordinary hacksaw. Typically, bunkerers focus their efforts on "manifolds," where multiple feeder pipelines are joined by bolts and welding that can be compromised with a certain amount of patience and the right equipment. To make sure a bunkering operation is successful, the valves must be opened to allow for maximum pressure, manifolds must be pried apart and welded back together quickly and cleanly, and everyone has to know that they're not going to be shot at by security guards in the process. From beginning to end, it's a process lubricated by official complicity and petty bribery. "No illegal bunkering would take place without the technical support of SPDC," says Sofiri Joab-Peterside, a sociologist at Port Harcourt's Centre for Advanced Social Studies, referring to Shell's subsidiary in Nigeria. "Most of these are [indigenous] contract staff, with no benefits, they are paid so little. They are so disgruntled that they become very willing allies." Before a ship is loaded with illegally bunkered crude, its owners will have "settled" the relevant security forces, the exact sum depending on the size of the vessel. For big barges, the Commanding Officer can expect as much as 2 million naira (about $15,000), the Base Intelligence Officer 1 million naira, and the Officer in Charge 500,000 naira.

In recent years, the federal government has made what appears to be a genuine effort to get illegal bunkering under control, but in doing so has only drawn attention to how entrenched the problem is at the highest echelons of the Nigerian political landscape. In late 2003 the *MT African Pride*, a Nigerian tanker vessel, was intercepted at sea near Shell's Forcados oil-export terminal, carrying some 11,000 barrels of unauthorized crude in its bunkers. Its thirteen Russian crew were arrested and the ship seized by the Nigerian navy, but several months later, in August 2004, it emerged that the *African Pride* had somehow disappeared from custody. For reasons unclear, the Russians had been allowed back on board mere days after their arrest, and the 11,000 barrels of crude had somehow been transferred to another ship and replaced in the *African Pride*'s bunkers with 11,000 barrels of seawater. Two of Nigeria's highest-ranking rear admirals were subsequently court-martialed and sacked for their role in the ship's disappearance. For many Nigerians high-profile cases such as this are proof that illegal bunkering has morphed from an activity enjoyed by frustrated youth looking for easy money into a professional industry managed by tightly organized, heavily armed syndicates of bunkering Mafiosi. At the very top of this Nigerian Cosa Nostra, it is widely believed, sit some of the country's most senior politicians.

Although ever-growing numbers of the Delta's residents are in some way involved with a bunkering mafia, few are willing to speak with foreign journalists and even fewer to serve as guides to those wanting to watch bunkering activities up close. After all, the sudden approach of a speedboat carrying a stranger with a notepad or a camera toward a manifold being pried apart by a group of young men with AK-47s, surrounded by jerry cans filled with highly flammable liquid, is unlikely to end happily. However, in Port Harcourt, "Nelson," a quiet and sincere young Ijaw from Oluasiri, was willing to see what he could do for me.

We met over a 7UP in the bar of my hotel, where he told me conventional crude oil bunkering was yesterday's news. The mafiosi who controlled it kept the lion's share of the profit and used the disenchanted local youth only for labor. "Not everyone could partake in it," Nelson said. "It was exclusively a game for rich people." In spring 2004 a more pervasive, and far more dangerous practice—which Nelson called "local bunkering"—had emerged.

In an effort to reclaim bunkering from the bunkerers, Nelson's community had turned their attention away from the crude oil and toward a lower-hanging fruit: natural gas. Enterprising youth had discovered the Shell gas pipeline going through Oluasiri, laid deep at the bottom of the riverbed. They had hired teams of divers to drill holes at three different points along the gas pipeline and attach hoses to the boreholes. Where the hoses came up to the surface, valves had been attached to control the flow of gas coming from the pipeline. The product that comes out of those valves is a volatile substance that is not quite crude oil and not quite kerosene (which is what crude turns into when the gas has been distilled out of it), but something in between that still has a lot of gas in it. The bunkerers let it sit for two or three days until it turns into kerosene, which is then sold to villagers for use as a cooking fuel. This bootleg kerosene is not as pure as what is sold legally, and when people put it in their stoves, it can explode and kill them. But it costs a fraction of what it otherwise would, and provides a handsome income for the unemployed youths of the area.

But the youths haven't been content to dash a bit of kerosene to the womenfolk, or distribute it in nearby villages. They have begun selling the product to black-market petroleum marketers, who take it to the larger provincial towns in the Delta such as Mbiama and Yenagoa. Some bunkered kerosene from Oluasiri has made it as far as Warri and even neighboring Cameroon. Many youths have also become involved in "trucking," a practice whereby tanker trucks half-filled with refined fuel stop off on their way to deliver to filling stations and top up their tankers with a little bunkered kerosene. The diluted product is then sold to unsuspecting gas stations for the price of regular unleaded.

"Crude oil is no longer marketable," said Nelson. "The federal government has done a lot to put a stop to that. So now we have this local bunkering. And unlike with crude oil, now *everybody* is involved. Shell staff, security guards, *everyone*." A note of despair was audible in his voice as he told me how an organized-crime racket had become a game for women and children. "Almost everybody has become a petrol dealer now. People come night and day to fill up their cans."

Nelson was a civil servant, working part-time toward a master's degree in social studies. In his research, he was looking at the social dynamics behind why people in the Delta have turned to illegal bunkering. But as time went on and he watched his less-educated friends get rich overnight, this budding academic was finding it harder to keep a bright line between student and subject. "My friends keep encouraging me to get involved, and I have been tempted." He looked down into his 7UP, clearly weighing how much more he wanted to say. "In fact," he said finally, his face filling with shame, "I am now also making some arrangements to join in."

Nelson explained how an operation's overhead costs break down. The divers are paid 1.5 million naira to dive down and connect a hose to the gas pipeline—generally, they have worked for Shell previously as contract laborers and laid the pipe themselves, so they know exactly where it is and can exact a steep fee for this expertise. Shell security operatives have to be bought off as well, so they will warn the bunkerers if soldiers or police approach as part of a crackdown. Another 1 million naira a month is typically given to the Shell employee responsible for operating the valves, so he will keep the pressure up on the pipeline. Just before Christmas, when pipeline pressure tends to be low, local contractors working for Shell had demanded 4 million naira from the community to keep the pressure up. The bunkerers had not been able to raise the funds, so Oluasiri had had a lean Christmas.

"If you are not corrupt, your family will curse you, your community will curse you, your wife will curse you," Nelson said with resignation, and more than a hint of self-justification. "Corruption is now the name of the game."

The really dangerous work, however, falls to the three or four area boys who "own" each of the loading points. After a while, the boys start to look pale and unhealthy from days spent inhaling natural gas at close quarters. Not to mention that the slightest spark—even if from two bits of metal rubbing against each other—can set off an almighty explosion that will be heard and seen for miles around. This happens with alarming frequency, and many young men have been killed stealing oil. In May 2006 more than two hundred people lost their lives when a pipeline outside Lagos burst into flames in the early-morning hours. Investigations showed hundreds of jerry cans lined up near the site of the explosion, an indication that locals had come under cover of night to collect fuel from the damaged pipeline.

But grisly stories like this don't deter anyone in the Delta. "Each of them is four million or five million naira richer than he would be otherwise," said Nelson, putting on his sociologist's hat. "How can you tell them to stop? How can you even talk to them? A civil servant makes 18,000 naira a month, and you have a young boy making 1 million naira a month. Today, even if you give these boys a free scholarship to go and study in America, they wouldn't go."

Nelson agreed to meet me the next morning for another trip into the creeks. He promised nothing, but said we could probably pay a discreet visit to Ijaw-kiri, an illicit trading post for stolen crude and gas.

⤙⤚

The trip to Ijaw-kiri began much as the trip to Kula had begun: with a long, bumpy drive. This time, though, it was to Ogbia, a backwater teeming with garbage and stray dogs. On the way we passed the town of Mbiama, a major entrepôt for the bunkering business. Along the riverbank, next to a bridge, a dozen tanker trucks were parked in a row, apparently unattended. At night, they would be loaded with bunkered fuel from speedboats.

Once in Ogbia, Nelson and I hired a decrepit-looking speedboat and clambered aboard. After an hour of darting and diving through the creeks, we arrived at Ijaw-kiri, which seemed the usual collection of a dozen straw and concrete huts, not unlike Kula. But in Kula, I had been greeted by proud village elders and curious children. Here, I landed to find a makeshift terrace with plastic tables and chairs, and something resembling a pinball machine, a rudimentary version of the contraptions seen on *The Price Is Right*. I was told it was called "lucky game" and that people played it for money. Four fat, sweating teenage girls giggled and beckoned me from behind a corner, calling out "*oyibo*" (a Nigerian nickname for white man). They were there to service the local lads who worked in the bunkering business, helping them to unload their newfound cash as quickly as they had acquired it. Ijaw-kiri was the Oluasiri area's very own Las Vegas—a little mud-and-wattle sin city buried deep in the jungle and owing its existence to the trade in stolen oil.

Nelson had me sit on the terrace while he went to negotiate on my behalf. A resentful-looking girl came up to my table and stared at me, until I realized she was waiting to take my order. I asked for a Fanta, and a warm orange soda was brought out. Just a few hundred yards away, and clearly visible, was the giant Soku gas terminal run by Shell, its exhaust stack spitting a tall orange flare. While I waited, a speedboat pulled up and five soldiers in combat fatigues clambered onto the jetty, rifles dangled carelessly over shoulders. They seemed uninterested in me, the nervous-looking *oyibo* with no good reason to be there, sipping Fanta on plastic lawn furniture. Instead, they ducked in and out of each hut, collecting their hush money, demanding soft drinks, and generally existing in a cloud of high-fives and abdomen-busting guffaws.

Nelson came back and said he wasn't having much luck. We moved to another concrete-and-tin shack a few yards away. More plastic lawn furniture, more rounds of Fanta. It became apparent that virtually no one lived in Ijaw-kiri and that every house was actually a bar. Eventually, a thin, withered-looking man emerged and, after a few words in Ijaw with Nelson, agreed to talk.

He was thirty-six, and a timber dealer by trade. In the dry season, when times were good, he could make as much as 100,000 naira (almost $800) a month. In the rainy season, he made nothing. By local standards, it was not bad money, enough to keep his wife and seven children from starving. But he thought his timber-dealing days were probably behind him now. In the past three or four weeks, he had become involved in a far more lucrative trade.

"Three years ago, I started seeing people improving their lives," he told me, speaking in Ijaw with Nelson translating. His lack of comfort in English, Nigeria's official language, was a sign that he had not had more than a few years of primary-school education. "I began to watch so I could learn the ins and outs." He explained that he worked in a team of ten to fifteen people, who together netted between 80,000 and 100,000 naira (around $600-$800) a day. On a good day, though, the figure could be ten times that—enough to give each member of the team as much money as a civil servant like Nelson would earn in a month. I asked the man what his wife thought of his new profession. He told me that when he had talked to her about it, she had simply said, "Anywhere there's money."

The bunkered fuel was loaded onto "zeeps"—giant containers that hold anywhere from 700 to 3,000 liters of fuel—and brought back to Ijaw-kiri, where each liter fetched around 9 naira (seven cents). From Ijaw-kiri, boys loaded the zeeps onto speedboats, and took them to Mbiama, Yenagoa, or Port Harcourt, where they could fetch as much as 45 naira (35 cents) a liter. Jerry cans were also sold in Ijaw-kiri for local consumption, a 25-liter jerry can going for a mere 500 naira (less than $4). He paused halfway through this explanation and looked imploringly into my eyes. "You know, I have to take care of my family."

When I visited Ijaw-kiri in mid-January, the pressure on the pipes was low, so people were out of work and idle. The place had a noticeably languid air to it. I thought the lull might make it possible to visit one of the three loading points (which had turned into five since Christmas), but it was made clear that this was out of the question. Each would be manned by a gang of teenage boys with assault rifles who would not take kindly to having their exact whereabouts noted by a foreign journalist.

On the way back, we passed a dozen stray zeeps bobbing along the water's edge. Some were white plastic, others rusted iron. Like floating tombstones, they were all that remained of boats that had been overladen, only to capsize and sink. Disks of thick black oil oozed out of them and into the creeks. Nelson informed me that even our speedboat was running on bunkered kerosene.

Back in Port Harcourt, a senior government official reluctantly confirmed that local bunkering had reached epidemic proportions in Ijaw country. He told me that just before Christmas, an explosion and fire had killed several bunkerers, and that he had approached some

Ijaw youths the next day to see if this tragic turn of events had given them any pause. "Quite the contrary," he said, "they were back at it the next day." The Christmas season had strained their finances, and they needed to go straight back to work to make ends meet.

It was also just before Christmas that Shell had laid off a thousand local employees as part of a budget-cutting exercise, and there was mounting concern that sacking skilled people in this environment might aggravate the situation. As Sofiri Joab-Peterside at the Centre for Advanced Social Studies pointed out, hundreds of technically skilled, disgruntled former Shell employees roaming the Delta with intimate knowledge of the company's operations and an obvious need to replace their lost incomes was a "terrifying scenario."

-<-->-

If illegal bunkering was just illegal bunkering, a Robin Hood case could almost be made for looking the other way. Whether it's the localized natural gas and kerosene trading engaged in by ordinary citizens, or the big-ticket organized crude oil trade managed by a well-connected elite, it could be argued that stealing crude oil from large multinational oil companies and redistributing the profits to the people most affected by their operations is no worse a crime than the wholesale looting of "official" state and federal oil revenue by senior politicians that has been going on in Nigeria for decades.

Unfortunately, nothing in the Niger Delta is ever as simple as it looks.

The really troubling thing about illegal bunkering is not so much that stolen oil is sold for profit, but what the profit is spent *on*. If the giggling prostitutes and free-flowing Fanta of Ijaw-kiri absorbed the bulk of the bunkering cash floating around the Delta, there might be some hope for the region. However, in a trend that has escalated rapidly since the late 1990s, much of the money from stolen oil has been spent on AK-47s and rocket-propelled grenade launchers—rough-and-ready hardware for a violent ideology of ethnic separatism. Local boys with no jobs, no access to schools, and almost certainly no real future ahead of them have been rounded up and organized into gangs and militias that inevitably clash with authorities and with one another.

During election campaigns, particularly those of 2003, many of these armed gangs have been hired by state and local politicians looking to intimidate communities into voting for them. Many expect the 2007 election season to see a return to the use of such armed militias by political campaigns. Leaving aside what it means for the future of a young democracy to have elections fought down the barrel of a gun, no one seems to have thought seriously about what the gangs would do after the elections were over. After the 2003 campaign, most Niger Delta militants felt their illegal activities—from bunkering to arms smuggling—had been given a veneer of legitimacy, not to mention impunity, by the powers that be. Some gangs became enlisted in the already fiercely divisive contests for village chieftancies—channeled into supporting one side or the other in the battle for the spoils of local power. But many also felt that they had brought certain politicians to power, and that these people owed them something in return—something more than a little cash and a condescending off-you-go. When such rewards were not forthcoming, they grew more resentful.

Had it not been for a rather breathtaking series of events in September 2004, though, the world might never have seen just how serious the problem of armed gangs in the Niger Delta had become. A fairly obscure youth militia calling itself the Niger Delta People's Volunteer Force blasted its way into the international headlines that month when its leader, the charismatic Mujahid Dokubo Asari, declared "all-out war" on the Nigerian state and threatened to shut down the country's crude-oil production. As part of what it called "Operation Locust Feast," the NDPVF demanded that all oil companies evacuate their personnel from the Delta, or prepare to engage in fully fledged armed combat.

There was more than a little irony to the posturing and the militancy, of course. Asari and the NDPVF virtually owed their existence to Rivers State Governor Peter Odili. In 2001, threatened by the growing success and moral legitimacy of Felix Tuodolo's Ijaw Youth Council, Odili had engineered a split in its leadership—pitting Asari against the original founders. Asari assumed control of the group, and in the 2003 elections proved to be a loyal ally of Odili, helping him be reelected. But Odili soon dropped Asari, who returned to illegal bunkering as a way to pay for weapons for his boys.

By August 2004, the NDPVF had begun clashing with a rival gang, the Niger Delta Vigilante, led by Ateke Tom and hired by Odili to keep Asari in check. Dozens of people were killed and Port Harcourt paralyzed by the violence. In September, aware that he had been completely abandoned by his onetime patron in the State House, Asari told his boys the battle was on. In the space of few weeks, a small youth militia that had started life as an illegal bunkering ring had transformed itself into a rebel movement claiming 2,000 fighters and bent on bringing the oil production of the world's seventh-largest producer to a standstill.

The reaction of the Nigerian government, the international oil companies, and the global petroleum markets was as predictable as it was swift. Shell immediately evacuated two hundred staff from Ijaw country. The price of oil spiked over $50 a barrel for the first time in history. And the Nigerian government dispatched helicopter gunships to Port Harcourt to shell NDPVF positions in the outlying village of Tombia. Two hundred forty people were reported missing and eye-witnesses spoke of the fighting as "something they had seen only during the Biafra war." President Olusegun Obasanjo found himself under intense pressure from an oil-hungry United States to bring the matter to a quick resolution and, incredibly, invited Asari to the presidential palace at Aso Rock, just outside Abuja.

To have been a fly on the wall at the Aso Rock meeting would have been to take a front-row seat at the near-total unraveling of Nigerian state and society. For here, behind closed doors, was a scene that made a mockery not only of the amour propre and sovereign self-regard of one of Africa's most important nations, but also of the African tradition of deference to one's elders and leaders. Here was Olusegun Obasanjo, sixty-nine years of age, three-time president of his country, active at the highest levels of Nigerian politics since the early 1970s, and a decorated Biafra War hero to boot, being lectured on history and politics by an Ijaw youth leader who days before had been crawling through the creeks with a leaf tied to his forehead to ward against evil spirits. "I could crush you," the president is said to have shouted at Asari at one point during the negotiations. Between them—according to press reports—sat a stern-faced American official making sure that no one got crushed and everyone knew the score.

And the score was Crazed Ethnic Militants 1, Federal Government of Nigeria 0. Obasanjo ordered Asari to sell the NDPVF's weapons to the state and desist from armed struggle in exchange for blanket amnesty and a promise that Asari and the NDPVF would not be targeted by Nigerian troops, as well as an undisclosed sum of money believed to be worth several million dollars (notionally described as payment for surrendered weapons). After days of being chased through the creeks by Nigerian soldiers, Asari and his boys were suddenly given a large amount of money and told they were free to go, as long as they agreed to behave themselves and stop threatening the oil supply to the outside world. It was an extraordinary piece of capitulation on the part of an African head of state with the international stature of Obasanjo, and one that he was unlikely to forget in a hurry.

The oil companies returned to work and the price of crude returned to its previous level. Many Ijaw privately felt that Asari had betrayed their cause, that he had been bought off by Obasanjo and the Americans in Abuja, or that he was simply an opportunist and a fanatic. But in a sign of the profound deficit in true leadership among the peoples of the Delta, publicly at least, Asari quickly became a liberation hero for the Ijaw. Moreover, the NDPVF had simply handed over a few AK-47s to the federal government in exchange for money that would no doubt be spent on more AK-47s, and gone back into the creeks to prepare for round two. The episode had put the authorities in a weak and exposed light, and Ijaw nationalist feeling only strengthened.

Obasanjo bided his time, no doubt sure it would not be long before Asari returned to his mischief. Sure enough, the casus belli the president was looking for came in September 2005, when in an interview with a local newspaper Asari called for the dissolution of Nigeria as a unitary state. Asari was swiftly arrested by federal marshals and taken to Abuja, where he was arraigned on five counts of treason against the Nigerian nation. Asari turned up to the courthouse in a defiant mood, wearing a white T-shirt that read "Self-determination and resource control: any means necessary," which he replaced, once in the courtroom, with a black NDPVF shirt.

When I caught up with Dokubo Asari in April 2005, it was during the lull in his battle against the federal government—just a few months after his Aso Rock meeting with Obasanjo and a few months before his arrest. He had come out of the creeks to take up residence in a palatial home in Port Harcourt, and was giving audiences to representatives of international media. Two conspicuously unarmed boys in a white air-conditioned van took me to Asari's home.

On arrival, I could see why so many Ijaw felt Asari had been bought off by Obasanjo. In the driveway of the compound sat two gleaming Lincoln Navigators, obviously kitted out with every available option and accessory. Two dozen boys sat in a circle in front of the SUVs while Asari lectured them in Ijaw. I was led inside, unnoticed by the warlord, and asked to wait in the frosty comfort of the house's reception lounge. I settled quickly into one of the overstuffed leather sofas, and became absorbed in the Will Smith movie that was showing on a cinema-sized plasma-screen TV. Asari, the eldest son of a high-court judge, often tells journalists that he is "proud to have been raised with a silver spoon in my mouth," and insists that all his wealth is inherited rather than the result of any devil's bargain with Obasanjo, or the ill-gotten gains of illegal bunkering. Whatever the truth, it was hard to escape the conclusion that his Port Harcourt headquarters was more like one of Mobutu's palaces than a combatant command bunker for the revolution.

Over the better part of an hour, various people drifted in and out of the lounge, ignoring both me and the Will Smith movie. Asari's voice could be heard outside, growing occasionally tetchy and sanctimonious, occasionally quiet and sincere. Eventually I walked into the adjoining NDPVF office to inquire when Asari might be ready to receive me. Apparently the fearless leader was not aware I was waiting for him, because when the young apparatchik went outside to check for me, I heard Asari respond with a petulant "Aww, why you have to fuck me up like this so late in the day?"

A few minutes later, Asari strode into the office wearing a bright orange Texas Longhorns jersey with a giant "4" on it. It struck me as a slightly lazy choice of outfit for a committed anti-imperialist who claims Nelson Mandela and Che Guevara as heroes and has in the past praised Osama bin Laden for standing up to the "arrogance of the West." To be fair to the commandante, though, I had turned up at his home unannounced, so I could hardly expect full combat fatigues or traditional African dress. Asari looked me up and down and I started to introduce myself. "You look like an Arab," he interrupted. "Are you a Jew?" I explained that I was Iranian by background, and he seemed vaguely reassured. He began telling me about his two trips to Iran and how he was once on very friendly terms with the Iranian embassy in Nigeria. "But they have all capitulated now." There was a distant disappointment in his voice.

I was dying to press him on what he meant by this, but could sense that our time together was limited and so made an effort to stay on topic. Why, I asked, had he launched Operation Locust Feast in September? "We were forced into it by the State," he replied curtly. Had the NDPVF really disarmed? "Yes, but getting arms back is very easy," he said dismissively. "If circumstances call for it, if the government decides to fight us, then we will rearm." This much, at least, was hard to argue with. Rumor had it the federal government had paid the NDPVF $1,000 for each AK-47 they had relinquished last year—though the weapon rarely costs more than $200 on the black market. It was practically an invitation to rearm.

Only when I put it to him that he had been bribed into quiescence by the Nigerian authorities did Asari come to life. "They bought me off?" he asked, leaning forward and locking eyes. "They bought me off? What moral right do you have? I am one of the most critical people of the government of Nigeria. People can go ahead and say whatever they like. They were not involved with me. We were doing the talking and doing the fighting." He grabbed a framed photograph off his desk and thrust it in my face. It showed him in the bush in full warlord pose—all lumpen grimace and low-slung AKs, complete with belts of ammunition crisscrossing his unbuttoned chest, Rambo-style. "Do you think it's easy to walk around carrying arms like this for nine months?"

We were interrupted by a few of the boys from outside, who urgently wanted Asari to settle a dispute over money. Asari invited them in and lapsed into an Ijaw monologue, punctuated by operatic flourishes in English, presumably for my benefit. "With everything I have, I will fight the Nigerian state until they come to their senses" was one that seemed particularly calculated to chill. And then, a moment later, "Tell the SSS*—send a message up to Abuja that if they try to kill me and they fail, it will be mayhem. Oh, the mayhem! I will no longer fight them in the creeks. I will fight them in Lagos and Abuja and Port Harcourt!"

After forty-five minutes of this, I must have looked fidgety, because Asari abruptly shifted gears and started telling the boys an anecdote peppered with enough English that I could follow along. It had to do with Governor Peter Odili—Asari's onetime patron now turned nemesis. During a recent drive through the Delta, the governor's car

---

*The Nigerian intelligence service.

had run low on fuel miles from the nearest legitimate filling station, forcing Odili's driver to buy fuel from one of the boys alongside the road. But everywhere the car pulled over, the governor was warned that he was about to buy what has become known in the Delta as "Asari fuel" and he had to drive sheepishly away to avoid the awkward publicity of being seen indirectly funding the activities of the NDPVF.

Asari rolled himself into a great ball of wheezing guffaws as he told this anecdote, slapping his fist on the table and letting his eyes fill with tears at every mention of "Asari fuel." The assembled boys chuckled politely, but seemed far less amused by the story than the great mujahid.

-<--->>

The anecdote might have been entirely apocryphal, and was not likely to be verified by anyone at the State House in Port Harcourt. But it seemed a little uncharitable for the boys not to be rolling on the floor with laughter along with their leader. After all, few would dispute the fact that in many parts of the Delta today anyone who runs out of fuel will be hard pressed to refill with anything other than illegally bunkered product. And when a state governor is forced to buy Asari fuel, you know the situation has reached a farcical extreme.

However, few of the Delta's disgruntled youth see anything remotely comical about the epidemic proportions the illegal bunkering trade has taken on. Instead, most consider it a deadly serious game of chicken with the Nigerian state—a desperate cri de coeur from a lost generation that sees no other way to claim its hydrocarbon birthright. Everywhere you go in the Delta, you will hear the same story. Whether Ijaw or Itsekiri, Ogoni or Edo, lean and angry youths will look you straight in the eye if you ask them about illegal bunkering and tell you, "Resource control begins here." It is a slogan for a generation.

"Resource control" is a thorny issue for Nigeria, one that threatens the very unity of the country. In a nation where many people see themselves as Efik or Ibibio first and Nigerian second, it is very difficult to tell someone who is sitting on top of billions of dollars of petroleum wealth that his windfall must be shared with two hundred other ethnic groups—especially when the immediate needs of his own community appear so pressing.

But to understand why "resource control" is such an emotional concept in the Delta, it is important to go back to the early years of the independent republic, just before the Biafra War. The first two constitutions of independent Nigeria (and the only ones to be freely negotiated by democratically elected representatives at constitutional conferences)—those of 1960 and 1963—divided the country into three regions (North, West, and East) and encouraged friendly competition among the regions based on their natural resources. For the North, this meant developing cotton and groundnut industries and the processing of hides; the West focused on cocoa, rubber, timber, and palm oil; and the East set about exploiting palm oil and petroleum. Each region was required to give the federal government 50 percent of royalties and rent from any mining activities it undertook, but was free to invest the rest as it saw fit. Nigeria was governed as a loose federation of three semiautonomous and largely self-sufficient regions, much as it had been by the British.

After the devastating experience of the Biafra War, however, everything changed. The military junta that had assumed power just as the war broke out, now rudely awakened to the dangers inherent in allowing three powerful regions to develop under its nose, set about creating a far more centralized state—one in which revenue distribution became a key function of federal, rather than regional, authorities. In 1969 Parliament passed the Petroleum Act, making all oil discovered in Nigeria the sole possession of the federal government, which would distribute revenue to the regions from a central fund. Then a presidential decree in 1975 raised the federal government's take of oil revenue from 50 percent to 80 percent, right in the middle of the boom years. Three years later, the Land Use Decree of 1978 gave all land-ownership rights to state governors (who were then military appointees), mandating that any land not already owned by the federal government would thenceforth be "vested" in the state governor for the benefit of "all Nigerians." Previously land had been owned communally with customary law determined by traditional rulers and clan chiefs. The decree meant that the state governor could now legally expropriate land for oil and mining concessions, and the affected communities had no legal right to question the entry of an oil company onto communal land. They would not even be entitled to compensation for the use of their land; any payments would go to the governor's office. Finally, in 1979, when the military ceded power back to a civilian government, its leaders made sure the country's new constitution enshrined the principle that natural resources from underground did not belong to the landowner. By the time the generals ended their decade-long stint in power, the federal oil grab was complete and etched in stone. The three regions had been replaced by twelve highly emasculated states* and any suggestion of "another Biafra" had been decisively crushed for years to come.

Nigeria's second experience with civilian rule was even more short-lived than the first. In 1983, after four short years, the elected government of Shehu Shagari was overthrown by a military junta, and one as indifferent to the Delta's grievances as any that had gone before. In 1982 Shagari had created a special derivation fund for oil-producing communities, drawn from a ring-fenced 1.5 percent of federal oil revenue. This money was promptly embezzled by state governors throughout the 1980s, and it was not until the Umuechem massacre of 1990 that the federal government realized that it had a problem it could no longer ignore. It set up the Oil and Mineral Producing Areas Development Commission (OMPADEC) to administer the 1.5 percent derivation fund, raising the percentage to 3 percent in 1992. OMPADEC was initially headed by Albert Horsfall, an

---

*Gradually increased to the current thirty-six states.

SSS agent who was accused by a government report of operating OMPADEC's $95 million budget like a private fiefdom. But in one of those extraordinary ironies that sees observers smacking their foreheads and exclaiming, "Only in Nigeria!" the man who wrote the report, Eric Opia, himself a politician close to President Sani Abacha, replaced Horsfall at the helm of OMPADEC and proceeded to embezzle $200 million before being removed for "gross financial misappropriations."

In 1999, when Nigeria made yet another transition to civilian rule, the newly elected president, Olusegun Obasanjo, replaced OMPADEC with the Niger Delta Development Corporation (NDDC) and increased the derivation fund for oil-producing communities to 13 percent (of which 15 percent would go directly to the NDDC, the remainder of its budget coming from oil-company contributions and the federal account). The NDDC has proven markedly less corrupt and incompetent than its predecessor, but remains accused of being a bloated and unnecessary bureaucracy.

As the problems with OMPADEC and the NDDC show, the Delta's difficulties cannot be spirited away with a simple injection of cash. Finances at the state and local level are run in a manner that encourages graft by obscuring how money gets spent. The governors tend to exert very tight control over state budgets. In some states, even a 20,000 naira ($150) expenditure requires the governor's signature. Most activists have discovered that in this operating environment, any oil revenue that *does* make it back into the Delta from the federal government quickly gets swallowed up by the voracious appetites of their "elected" leaders in state and local government. These politicians spend it on ostentatious white-elephant projects, waste it on official cars and perks of office or, worse, salt it away into foreign bank accounts.

With few exceptions, Nigeria's state governors are holdovers from the period of military rule—political appointees who were selected for popular approval in the highly flawed national elections of

1999 and 2003. Though they initially remained loyal to their political bosses in Abuja, many have now formed their own power bases (often using militia support) and cultivate their independence from the national leadership of the ruling People's Democratic Party. Given the obvious material benefits to them, the governors from the nine oil-producing states have become overnight converts to the cause of "resource control" and are among the most vocal proponents of an increase in the federal derivation fund from the current 13 percent to something reflecting the 50 percent of oil and mining revenues that the regions were allowed to keep in the 1960s.

Just how far the Delta's governors were willing to go in defending their newfound principles became apparent in early 2005, when, in an effort to resolve many of the issues left unaddressed by the transition to democracy in 1999, President Obasanjo convened a special four-month "national dialogue" in Abuja to which he invited stakeholders from all sectors and regions of Nigerian society. Activists had been calling for years for a "Sovereign National Conference," a sort of first-principles meeting to reconsider both the constitution and the very idea of Nigeria as a unified nation. Obasanjo's "Confab," as it was called, was a compromise measure—there was to be no talk of breaking up the country, but all other ideas could be put on the table. But the Confab immediately ran into a minefield over the question of resource control, with delegates from the southern oil states insisting on a return to the original 50 percent derivation principle of the 1960s. When a suggestion of 18 percent was put forward, the southern delegates, led by the governors, countered with a "final offer" of 25 percent, then walked out of the meeting during its final days when their demand was not met.

For many of the Delta's delegates, 25 percent represented a genuine line in the sand, a figure that could not be lowered without bringing dishonor to the years of violent struggle under the banner of "resource control" that had claimed the lives of so many of their bravest young men. They were grateful for the heavyweight support of their governors, but few were naïve enough to believe it represented anything other than naked opportunism. As Patterson Ogon, one of the original activists of the Ijaw Youth Council, put it to me as we sat on the floor of his house one sticky Saturday morning, watching his family sweat through a fourth straight day without electricity, "Those people never wanted democracy. They tried to stop it. And now they are reaping the benefits."

When it comes to corrupt governors reaping the benefits, few can claim the level of expertise of Governor Diepreye Alamieyesegha of Bayelsa State. London is a favorite destination for looted Nigerian cash; some of the most exclusive addresses in Kensington, Knightsbridge, and Mayfair can be traced to members of the Nigerian political elite. So few were tremendously surprised when the Metropolitan Police arrested Alamieyesegha in September 2005, during one of his frequent visits to London, and charged him with three counts of money laundering. Alamieyesegha was already suspected by British and Nigerian authorities of stealing millions of dollars in cash from Bayelsa and using it to buy an oil refinery in Ecuador as well as several homes in London, California, and South Africa. A British judge released the governor on bail, but forced him to surrender his passport, in order to prevent him from returning to Nigeria, where acting governors are immune from prosecution. Weeks later, however, Alamieyesegha shocked both British authorities and most of Nigeria when he turned up in Yenagoa, the Bayelsa State capital, and told a crowd of cheering supporters, "I cannot tell you how I was brought here. It is a mystery. All the glory goes to God." According to press reports, Alamieyesegha had procured a fake passport and slipped himself through Heathrow Airport dressed as a woman.

Alamieyesegha was later impeached, but the entire saga is emblematic of the Niger Delta conundrum. Bayelsa is one of the smallest of Nigeria's thirty-six states as well as one of its newest, but it produces 30 percent of the country's oil. Its former governor is, to put it delicately, someone with a not-inconsiderable international-investment portfolio, who during his time in office treated Bayelsa as a contiguous part of his global real-estate empire. In 2005 the state budget set aside $8.5 million to construct official residences for the governor and his deputy, along with more than $2 million for furnishings and luxury appointments. All this is in addition to the $25 million already spent on the governor's mansion since 2002. (The fence alone cost $5.7 million.) Alamieyesegha was once a close ally of President Obasanjo and a loyal PDP member. But in recent years he has grown closer to Obasanjo's vice president, Atiku Abubakar, with whom Obasanjo has been falling out openly, and many believe Alamieyesegha's legal troubles are part of a mounting assault on Atiku's power base by the president. Obasanjo was so incensed by Alamieyesegha's disappearing act at Heathrow that he wrote a personal letter to Prime Minister Tony Blair asking for an explanation of the fiasco. But these millions have not been enough for Governor Alam. So there he was at the National Confab in April 2005, sticking up for the people of Bayelsa, and making a "principled" stand for his 25 percent.

As of late 2006, 33 of Nigeria's 36 state governors were under investigation for corruption, money laundering, or other financial crimes. Local governments, controlled almost entirely by the state governors, are merely another stop along the gravy train. And as for the tribal chiefs and traditional rulers, they are renowned for spending their oil company handouts and "compensation" awards on the "five Gs": guns, girls, gold, gin, and grass. After years of watching their patrimony squandered in this way, a large percentage of the Delta's population feels abandoned by both national and local politicians, and has settled on illegal bunkering as the most direct way to ensure that they benefit from their own oil wealth. The trouble is that what started as activism has become an industry. In the words of one activist, "It is becoming increasingly difficult to separate greed from grievance." Moreover, what started as an act of political defiance has helped entrench the powers that be. Profits from the trade in looted crude have either amassed to corrupt politicians from all over Nigeria, or been used to buy weapons and form the militias that are exploited by corrupt local politicians and traditional rulers desperate to keep themselves in power (and close to the opportunities for self-enrichment).

Given this steady erosion in the ability of the people of the Niger Delta to control their own resources, it is perhaps understandable that the 1960s have taken on a mythical status among Delta activists. So revered have these years become, so heady the nostalgia for the days before centralized control of petroleum resources, that many activists now insist that *genuine* resource control can never be achieved under the present system—that only a return to the "true federalism" practiced by the founding fathers of the nation (and before that by the British colonial masters) can guarantee a fair distribution of oil wealth for the people of the Delta.

This genuflection at the altar of "true federalism" has taken on the status of a sort of unofficial religion in the Delta in recent years—a precarious memory to begin with, elevated into a political rallying cry and romanticized to within an inch of its life. The exact extent to which Nigeria can be said to have been a "true" federation is subject to debate, especially among northern and western Nigerians who have far less to gain from resource control. Many in the North question bitterly the analogy made to the way they are permitted to keep the profits from their cotton industry, pointing out that it takes effort to grow cotton, whereas there is no effort involved in inviting foreign oil companies to extract petroleum. They argue that natural resources belong to the nation as a whole, that affected communities should be compensated for environmental damage caused by oil exploration, but not receive any special derivation funds beyond that compensation.

"What is true federalism?" asked one northern intellectual and businessman I met in Lagos. "Ask these people what true federalism would look like in the Constitution and they can't tell you. Is it just about resource control? All the oil in Norway is in the North Sea.

Does the North Sea region control all the resources? Is that how it works? Even the British didn't operate resources that way. How do you think Lagos was built? You see, there is this romanticization of the First Republic, of something that never existed. No, sorry, true federalism is a banal and superficial argument. You don't build a country based on one principle, which is the sharing of money."

This appeal to national unity and the fundamentals of nation building is common in rebuttals to southern arguments about true federalism. Especially in the North, where bloody intercommunal and interfaith rioting has often broken out in response to efforts by some state legislatures to impose strict Islamic sharia law, the experience of federalism has been largely negative. Many feel that the sharia crisis is a perfect demonstration of how Nigeria's real problem is actually an *excess* of federalism. They argue that in a more tightly centralized federation, states would not have the power to unleash ethnic hostility or spark religious tensions in that way.

And ultimately, as both the international oil companies and the Nigerian government are fond of pointing out, all the rhetoric over resource control and true federalism ignores the fact that Nigeria's enormous population means the nation is not quite as oil-rich as everyone thinks. Oil brings more than $10 billion into the Nigerian treasury every year, but divided between 130 million people, that is less than 25 cents per person per day. Even if 100 percent of oil revenue stayed in the Delta—an extremist proposition put forward only by the Asaris of this world—it would not make anybody rich overnight, except possibly the state and local politicians in control of the windfall. In the end, most sensible observers feel the problem is not simply resource control, but *equitable* distribution of resource revenues. Today in Nigeria, 80 percent of oil and gas revenue accrues to just 1 percent of the population. What this means in real terms is that virtually everybody in the Delta scrambles to get by in shantytowns built of driftwood and corrugated zinc, watching their children die of preventable diseases, while their corrupt leaders whiz past behind the tinted windows of air-conditioned BMWs.

-<-<->-

"The Niger Delta man, the Ogoni man, the Ijaw man, is as far from the Yoruba man or the Hausa man culturally, linguistically, and even physically, as Spain is from Norway, or as Portugal is from England."

I was back in Lagos. After spending weeks in the Delta looking for the face of "trouble," I had decided to cross back over the wavy lines into the cocoon of peace and tranquillity I had left behind. Not that anyone in their right mind would ordinarily describe Lagos that way. One of the world's largest cities with a population of at least 15 million people (experts have long ago given up on an accurate census), Lagos is a noisy and thickly polluted diorama of struggle and desperation in which even the dogs seem mildly possessed. It is a place so anarchic that it is not unheard-of for the police force and army to stage deadly street battles over who has the right to extort bribes in a particular neighborhood.

But if you've just come from the Delta, there seems a pleasing predictability to the madness here—a reassuring tendency for disputes to collapse in on themselves rather than ripple out across creeks and communities; the probably universal ability of city dwellers to shrug off the screaming disintegration of daily life.

I had dropped in on a conference of social and environmental NGOs at the Lagos Airport Hotel, where I had had the good fortune to bump into an old man whose heart was full. He was Alfred Ilenre, an active and well-respected Edo elder and the head of EMIROAF, the Ethnic and Minority Rights Organization of Africa, and he was giving me his version of Nigeria 101. Dressed in a long, unadorned white *agbada* that swallowed up his sinewy features, Ilenre was doing his best to speak slowly and choose his words carefully, so that even a Western naïf like me could understand.

"So you see," he croaked magisterially, "the problem is that the only thing the federal government ever did was to give the oil companies a map of Nigeria and say, 'Go and find oil there.' In the local communities, they knew nothing of this. All they knew is that one day they see a white man come. They see him come with three black men, and start digging. And they ask them what they are doing, and the white man shows them a piece of paper from Abuja. This piece of paper means nothing to them, but the white man says it's a piece of paper that says he is allowed to dig in the backyard of my house. It says nothing about whether this might be our ancestral home.

"So of course, inevitably, the people get angry. Lagos and Abuja could be New York or London to these people. You have in the Niger Delta people who are illiterate, who have lived eighty years and never even been to Lagos. Some have never even been more than five miles from their village, or as far as Port Harcourt. And all they know is that you're coming with technology to distort their peace, their serenity, and their survival. And Shell says to them, 'If you want compensation, go to Lagos, go to Abuja.' But these people don't know Lagos, they don't know Abuja. So they hold *you* responsible. They say, 'White man, we don't know what is happening, but we hold you responsible.'"

I felt the time had come to visit the white man, and to let the white man have his say.

◄─◄─►►

The e-mail, when it finally arrived, came from Shell headquarters in London. It informed me that Chris Finlayson, chairman of Shell's Africa division, had agreed to meet me on Monday at 9:00 A.M., at his office on the top floor of Shell's imposing high-rise on the Lagos waterfront.

Anywhere else in the world, this would have been perfectly reasonable, particularly coming from the top executive at a large corporation. In Lagos, however, it was a sadistic request. Many cities suffer severe and perpetual gridlock, but in Lagos the inability of traffic to move more than a few inches per hour during rush hour has reached comical proportions. Throughout my time in Lagos, morning appointments with Nigerians had been scheduled for 10:00 or 11:00 A.M., with a tacit understanding that arriving at any point before lunch was acceptable. But it was London calling now, and Monday 9:00 A.M. it would be.

I had been employing a "fixer" in Lagos—a BBC freelancer named Sam Olukoya, who in turn had arranged for a young man with a car to drive us from appointment to appointment. Until now, our days had got off to a late start, as the driver would have to drive across Lagos to pick up Sam before driving down to Victoria Island to meet me at my hotel. They always spent a couple of hours in morning traffic before arriving, late, hot, and frustrated. When I stressed to Sam how important it was that we be on time for my Monday-morning appointment with Chris Finlayson, he sucked his teeth and sighed. The only way to make that happen, he said, would be to have the driver sleep at his house the night before. They would set off together at 6:00 A.M. They should arrive comfortably at my hotel by 8:30 A.M., leaving plenty of time for the ten-minute drive to the waterfront. It seemed an extreme solution, but Sam was insistent. He did not want to let me down.

Monday morning 8:30 came and, just as I had feared would be the case, I was standing outside my hotel sweating through my suit. With my right thumb, I had perfected a pattern of button touches on my cell phone that was letting me redial Sam's number seven times a minute. Of course, there was no way I was going to get through, as everyone else in Lagos with somewhere to be would be doing the same thing. At the best of times, Nigeria's shiny new mobile-phone networks are a hit-or-miss affair, but at peak times, the country reverts to the state of communication paralysis it was in until five years ago, when asking for an appointment with someone meant getting a member of your staff—if you were lucky enough to have a staff—to drive across town with your diary and find out when the other person might be free.

Finally, at 8:55 A.M., the car honked from a distance and I ran over and climbed in. "There was an accident on the bridge." Sam sounded genuinely crestfallen by the spectacular failure of his best-laid plans. "We have been on the road since before 6:00 A.M." And looking at the thousands of cars squeezed together on the expressway like pieces of a snugly fitting jigsaw puzzle, I believed him.

Suddenly—and without a hint of aggression on his placid adolescent face—our driver nudged several motorcyclists with his bumper to clear a wedge of space between two trucks loaded with hay; honked and pushed his way through the roundabout, making a goat and several women balancing plates of oranges on their heads jump out of the way; and tore off on an open path toward the waterfront, delivering us, miraculously, in front of Shell HQ at 9:00 A.M. on the dot.

My relief at arriving on time dissipated quickly when I realized how many layers of security I would have to go through. I suppose I should have known that one doesn't pull up with a screeching halt in a clapped-out old Mercedes billowing exhaust and run out in a sweaty huff demanding to see the chairman of Shell's Africa division. But the ballet of sign-in sheets and swipe cards I was asked to perform was a parody of procedural overenthusiasm. Finally, at 9:10 A.M., as I sat in the second of two entrance lobbies, a man emerged from the elevator to take me to the public-relations suite. After a byzantine ride that had us going up to the sixteenth floor and then back down two flights by a staircase and around to a separate wing of the building, I was dropped off in an open-plan office where several Nigerian staff were sitting quietly at their desks.

One of them looked up at me, then looked sternly at his watch.

"Are you Mr. John?" he asked sourly.

"Yes, that's me," I puffed impatiently, trying to sound important enough that I might be able to dispense with formalities and be led straight to Finlayson's office.

"You have kept me waiting," he said.

"Yes, I'm terribly sorry." I began rehearsing the story about the accident on the bridge, annoyed that the PR flack was making me later by the minute.

"Well, that's all right, you are here now," came the curt reply. "So tell me, how can I be of help?"

I was thrown by the question, which sounded like an invitation to start an interview. Surely this Nigerian man in the open-plan office wasn't . . . ? No.

"Um, I'm here for my appointment with Chris Finlayson," I said.

"Yes, I am Chris Finlayson," he replied impatiently. "What can I do for you?"

"Oh! You're . . . Oh, um. . . ."

And just like that, it was all over. Even before it had begun. Sure, the interview would probably still go ahead, and I would run through my list of lovingly prepared questions. But the bottom line was that I had turned up fifteen minutes late, and immediately let slip a giant, silent racist fart. Anything I said now would only draw more attention to it. I had to pretend it had never happened, and hope no one would notice. After all, I had known since I was eight that whoever smelt it dealt it.

As I stood there frozen, trying to think of the right face-saving platitude, a mischievous grin spread slowly and evenly across the black face my mind had begun to accept was Chris Finlayson's. The deep, dark eyes twinkled. "I'll take you up to his office," he said, as he handed me his business card and slapped me gently on the shoulder. *Bisi Ojediran, Head of Corporate Communication and Media.* I could hear the woman behind me desperately suppressing a chuckle.

By the time I walked into Chris Finlayson's office, it was 9:20 A.M. and the crisp gray suit I had put on that morning was a blotched body glove of perspiration. Between the Darwinian exercise of Lagos rush-hour traffic, the carnivalesque maze of lobbies and handlers and bar codes of Shell Tower, and the cognitive ambush of Bisi Ojediran,

I was in a state of disheveled apology and near-total disarmament—my weapons surrendered, my troops demobilized, and my generals ready to sue for peace on almost any terms. As my shoes sank into the plush pink carpet of Finlayson's office, as my sweat glands contracted sharply in the bracing air-conditioning, and as I glanced out of the large windows at a sweeping panorama of Lagos, I registered a lesson that would come to serve me well in Africa: The best kind of psychological warfare is the kind you never saw coming.

As if to reinforce the point, Finlayson was faultlessly charming and not the least bit irked by my late arrival. Heavyset and ponderous, with a bushy goatee, Finlayson chose his words carefully and gave the impression of someone who genuinely believed them. He began by telling me with some pride that Shell had recently moved its entire Africa staff from The Hague to Africa, "on the principle that we want to run Africa *from* Africa." We then talked a little about Shell's future in Nigeria from a commercial perspective, and he spoke in excited terms about the company's offshore Bonga concession, "an area with great promise." I wanted to take this as a subtle admission that Nigerian *on*shore drilling was becoming too dangerous for Shell to depend on in the long term, but this was probably unfair—at 225,000 barrels a day, Bonga was clearly going to be one of the company's biggest new developments.

We turned next to the thornier questions of Niger Delta politics, and Shell's role in stoking the anger of the communities in which it operated. Here, Finlayson was anxious to paint a picture of slow-but-steady improvement. "It is true that, up to the period of the return of democracy [1999], the Niger Delta has certainly been starved of government funding," he admitted. "But this is a position that changed six years ago. . . . If you look at the new constitution, if you look at the funding that goes to people in the Niger Delta through government on a per capita basis, it's now ten times what a person in Kano State [in the North] receives. So this whole—as they call it here—'resource control' debate really very much comes down to the fundamental questions now of the existence of Nigeria. How much of the revenues should be kept for the oil-producing states and how much are justifiably spent in the rest of the country?"

It was the standard argument made by Nigerian politicians in the face of demands for greater resource control in the Delta: If you say the percentage of money from oil exploration that we are letting you keep is not enough, then you are forcing us to question your commitment to Nigeria as a unified nation. If you accept that you are Nigerian, if you claim you are not looking for another Biafra War, then you must be prepared to share the wealth with your fellow citizens, wherever in Nigeria they live. Past a certain point, carping about "resource control" and "true federalism" becomes irreconcilable with Nigerian patriotism.

Next, Finlayson ran through some of the international oil industry's favorite facts and figures, chief among them that 70 percent of Shell Nigeria's employees are from the Niger Delta, that 90 percent of total exploration and production revenues accrue to the Nigerian government (by far the highest percentage in Africa), and that in addition to their corporate taxes and an $80 million annual contribution to the NDDC, Shell gives a significant amount to community development projects every year. He insisted that to do any more would be to tread on the toes of the Nigerian government and its responsibility to its citizens.

"To what extent should an international oil company make the decisions about development aid for the government?" he asked, adding quickly, "This isn't simply saying, 'It's not our fault' and standing away from it. But there is a genuine moral dilemma there. Should we set ourselves in the position of saying we know better or we have more democratic right than the government does to decide who gets development aid? That's the balance. In the end, the primary agent of development in any government should be the government—particularly, I would suggest, as you come into a democratic government with reasonable legitimacy, with a high proportion of the revenues. Rather than saying, 'We'll up our costs so we can give more money out directly,' we're saying, 'No, our duty is to minimize our costs, meet our social obligations to the community, pay the maximum tax to the government under the terms of the treaty, and then it is up to the elected representatives of the people to decide how to spend the money.'"

This is an argument for which NGOs and activists have little patience. They point out that the oil industry is the mainstay of the Nigerian economy, and that a highly symbiotic relationship exists between oil companies—especially Shell—and Nigeria's politicians. For decade after decade, Nigeria was ruled by a succession of military regimes, each of which came to power by overthrowing its predecessor in a coup, and many of which treated the Nigerian state's contracts with multinational oil companies as a license to print money for themselves and their families. Foreign oil companies operating in Nigeria went out of their way not to rock the boat, cozying up to corrupt dictators and asking for—and getting—brutal military protection every time they felt their facilities were threatened by protesters. For years, the closest thing they had to a long-term strategy of community relations was an unofficial policy of paying off those communities to let them work in peace. NGOs and activists argue that the oil industry possesses the ability to influence government policy and should not shy away from the country's problems. Shell and other multinationals counter that, as individual companies, they cannot pressure government too much without risking losing out to other companies on competitive concession rights; and that even if Western companies presented a unified front in favor of a better deal for the Delta, there is nothing to prevent such concessions going to, say, the Chinese.

Finlayson also insisted that Shell was no longer in the business of paying bribes to communities in exchange for a peaceful working environment. "We have clear rules now that say no cash payments to communities unless there is a genuine service that is provided and no so-called ghost workers. Those have been practices that have taken place in the past. We've stopped them. It's tough stopping them, but we're bringing that in right across the industry."

I asked him about the recent problems with Kula, a community that had never really given Shell problems in the past, and asked why he thought it was, if things were improving in the Delta, that Kula had started acting up. He admitted, "With over eight hundred communities, you cannot personally look after each one. The risk is that you look only at the communities that give you problems." Given the industry's history with ghost jobs and hush money, a cynic might conclude the problem with Kula was that no one had bothered to buy its quiescence recently.

My final question for Finlayson had to do with Shell's future in Nigeria. The company had recently commissioned a confidential risk-assessment survey from a consulting firm called WAC Global Services, and WAC's final report had been leaked to the press, much to the embarrassment of Shell. Among other things, the WAC report concluded that it was "clear" that Shell "is part of Niger Delta conflict dynamics and that its social license to operate is fast eroding." WAC warned that if things did not improve, Shell would not be able to operate in the Delta "beyond 2008" without violating its own business principles. The suggestion that Royal Dutch Shell, one of the world's largest oil companies, might have to beat a hasty and ignominious retreat from one of its most lucrative and long-standing reserves, had arched the eyebrows of petroleum industry analysts around the world.

Finlayson said he actually agreed with 95 percent of what was in the report, but that he respectfully disagreed with the report's overall tone of pessimism and saw leaving Nigeria as a sort of doomsday scenario, conceivable but ultimately unlikely. Reading the ninety-three-page WAC report, it is hard to see exactly which 5 percent Finlayson might have disagreed with, as it is a wall-to-wall catalog of bad news, dire predictions, and damning condemnations of the company's past business practices in the Niger Delta. Among other things, it accuses

Shell of a "quick-fix, reactive, and divisive approach to community engagement," and describes the company's conflict-management initiatives as "limited in scope, under-resourced, [and] undermined by lack of coordination, coherence and analysis." It finds "a lucrative political economy of war in the region" that it says is "worsening and will deeply entrench conflicts."

Finlayson's response seemed a quiet acknowledgment that there was a limit to how much even a topflight oil executive could spin a bad situation. In fact, during the time I was in Nigeria, it was possible to detect in several of the supermajors a growing willingness to say, "Look, we know we've made a lot of mistakes and we've been burned; but, honestly, at this point we're not quite sure what we can do to make this thing work." There was a certain bluntness, a modest attempt at transparency where once there had been only defensiveness and secrecy. Shell had not only made Finlayson available to me on short notice, but had also spent a day flying me around in one of its expensive helicopters, paying drop-in visits to the Bonny and Soku facilities and a community-agriculture center that it was funding in Ogoniland. Chevron, meanwhile, had taken out two-page ads in Nigerian newspapers, acknowledging in rather unambiguous language the sins of the past and announcing a new approach to community relations—a so-called "Global MoU" to supersede all previous MoUs signed with host communities. The divisive concept of "host communities" was to be scrapped, in fact, and replaced by more formalized and geographically inclusive "Regional Development Councils." In an impressive act of corporate humility, Chevron admitted in the ads that its approach to the Delta in the past had been "inadequate, expensive, and divisive."

But it was hard to escape the conclusion that the mea culpas and the self-flagellations might all be too little, too late. There are people of genuine goodwill working for the oil companies, many of whom are determined to find a way to earn that elusive and much talked-about "SLO" (social license to operate) that makes it possible for their employees to work in peace in the Delta. But increasingly, alongside good intentions and innovative approaches to community and media relations, there is also a mood of resignation and gathering despair, a feeling that things may have gone too far, too deep, in a way that no amount of goodwill can turn around. "The Pope himself could not fix things now" was the way one activist described the situation of the Delta to me. "He would just be corrupted or killed or co-opted by one group or another. Today, every little boy in Nigeria is talking about 'big money,' not hard work. People are assassinating one another to become local councillors. How can you turn something like that around?"

And indeed, throughout 2005 and 2006, the situation in the Niger Delta continued to deteriorate. After Asari's arrest in late 2005, the NDPVF appeared to lose momentum. But it was quickly replaced by another Ijaw group, the Movement for the Emancipation of the Niger Delta (MEND), that demanded the immediate release of Asari as well as Governor Alamaseyeiegha (who by then had been impeached and rearrested by allies of Obasanjo). MEND appeared to incorporate many former NDPVF fighters and swore to be even more ruthless and uncompromising than its predecessors. In January 2006 MEND burst into the headlines when it kidnapped four Shell employees and held them for nineteen days before releasing them on "humanitarian grounds." In February nine oil workers were kidnapped in the Delta and a crude-oil pipeline owned and operated by Shell blown up. In May eight more hostages, most of them American, British, and Canadian, were taken and released quickly, as were five South Koreans a few days later. In August alone, nineteen foreigners were kidnapped. By the end of 2006, the number of oil workers taken hostage in the Delta stood at over seventy, a grim new record for the region.

In one particularly shocking incident in August 2006, gunmen burst into a nightclub popular with expats and began shooting into the air. Four foreigners were abducted in front of bewildered security guards and driven away. As the year wore on and the 2007 election season got into full swing, it became increasingly hard to maintain the illusion that the Delta was a safe place for oil companies to send their employees. Expatriates in Port Harcourt began commuting to work in the mornings with armed police escorts, and placemats at a bar popular with expats carried the advice: "Eat a lot. Fat people are harder to kidnap."

Instances of pipeline vandalism also increased. In October 2005 a pipeline fire in Delta State killed about sixty people. In December armed men in speedboats dynamited a Shell pipeline in the Opobo Channel. In January 2006 a pipeline attack from the Brass Creek fields to the Forcados terminal forced Shell to cancel its delivery commitments to the end of February. Additional attacks in February extended the force majeure indefinitely. By October 2006, the little kingdom of Kula that I had visited had grown restive again, and taken over the local flow station, just as they had promised me they would. Throughout most of 2006, in fact, some 600,000 barrels of oil a day—25 percent of Nigeria's output—was shut in, thanks to militant activity in the Delta.

And all this in a country that once had so much hope, a country that oil was going to make powerful and prosperous beyond its wildest dreams. The story was never supposed to end this way.

In fact, it has now become apparent to just about everyone in Nigeria that oil, far from delivering Nigeria's citizens into a Valhalla of wealth and prosperity, far from creating a sort of African Saudi Arabia where the world's poorest and most dispossessed are catapulted into a whizzy tomorrowland of roads and factories and spotless hospitals, has brought only a spectacular and grotesque form of wealth and ostentation for a lucky few and a steady decline in the overall health of the nation's economy. In 1960, when the British handed over the keys to a new generation of indigenous idealists, the country was nearly self-sufficient in food, and agriculture accounted for 97 percent of export revenue. Whether it was cocoa or yams or groundnuts, Nigeria was an African breadbasket, feeding both its citizens and those of other countries. And when the oil boom kicked in, it seemed that there was no stopping Nigeria. From 1965 to 1975, the federal government's annual revenue spiked nearly tenfold, from $295 million to $2.5 billion, with 82 percent of that new wealth accounted for by oil—an astonishing economic miracle that, on paper at least, made Nigeria the world's thirtieth-richest country and should have been the cause for national celebration.

Instead, as the oil boom stuffed the government's coffers, the country's agricultural base saw the bottom fall out. From 1970 to 1982, production of cocoa fell 43 percent, that of rubber fell 29 percent, cotton 65 percent, and groundnuts 64 percent. Nigeria was quickly ensnared in a seemingly endless cycle of national debt, which was accompanied by a spectacular collapse in the living standards of its citizens. The percentage of Nigerians living in poverty went from 28 percent in 1980 to 66 percent in 1996. Average annual income, which in 1980 was $800 per person, today stands at a mere $300. And of course, throughout these years, the seeds of discontent and anger were sown in the Delta—seeds that by the 1990s and early 2000s had sprouted into a thick forest of murder, kidnapping, organized crude-oil smuggling, and violent gang warfare.

On the cultural and political front, it has been much the same story. In the 1970s Nigerians came to see their country as a budding African powerhouse, an attitude that has proven resilient and led many neighboring countries to resent Nigerians for their perceived arrogance and sense of superiority. Nigeria became a champion of the antiapartheid cause, and hosted the Festival of African Arts and Culture (Festac) in 1977. A booming middle class found it could afford cars and televisions for the first time, and start up businesses. There was even talk of Nigeria developing nuclear power and creating an African high command.

By the 1990s, however, Nigeria had become just another African basket case, its fall from grace rendered all the more dramatic by its onetime ambitions. Throughout the world, Nigerian travelers were treated with suspicion and contempt, assumed to be drug runners and money launderers, and subjected to humiliating searches and questioning at airports. Signs went up in security lounges in the United States warning international air travelers that the Lagos airport did not meet basic safety and security standards. The venal regime of General Abacha, who salted away billions in foreign banks and had people executed for their environmental activism, drew scorn and ridicule in all corners, until the man himself finally expired in 1998—fittingly enough, in the arms of two Indian prostitutes.

Talk to the average Nigerian, and he will be quick to blame the country's leaders for this fiasco, convinced that if only the country had been a bit luckier, if only she had been blessed with "better leadership"—the kind that knew how to manage an oil windfall for the benefit of all—then things might have turned out very differently.

Talk to economists, and you will hear a very different message. They will tell you that what has happened in Nigeria is the result of an almost-insurmountable confluence of structural changes that nearly always accompanies a resource boom in a Third World country. They will remind you that there is a name for this phenomenon. It is called the "curse of oil," and it can take many forms. In Nigeria, where most oil exploration takes place onshore, the oil curse has taken the form of dramatic political violence and hostility among communities, oil companies, and government authorities, making oil companies keen to explore Africa's more-peaceful offshore potential. However, as we will see in coming chapters, the curse of oil is a hydra-headed monster, and can have far more subtle sociological and economic manifestations. And as a dozen or so African countries begin to experience little oil booms of their own, they will be watching closely the experience of the behemoth Nigeria, and asking themselves if they, too, will be so unlucky.

CHAPTER 2

THE OFFSHORE ILLUSION

IF YOU BELIEVE everything you read in the press, you might be forgiven for thinking that Africa never had oil before; that it has rained down on the continent, like so much manna from heaven, just in time to keep the price at the pump from going north of $3 a gallon in Fargo.

In reality, sub-Saharan Africa has been supplying a healthy flow of crude oil to the international market for decades. Nigeria made its first shipments of oil in 1958, two years before it had even declared independence from Britain, and the lush tropical forests of Central Africa have been drilled by French companies since the early 1950s.

In fact, it was in the 1970s, when the Arab oil embargo gave the world its first look at what an uncertain supply of crude might be like, that oil companies began looking seriously at Africa's unexplored hinterland as a possible source of new oil. Exploration wells were drilled in virtually every part of Africa as entrepreneurs combed the jungles and savannahs with the resolve of Victorian explorers seeking the headwaters of the Nile.

Oddly, oil was found nearly everywhere in Africa that holes were drilled, but nothing much ever came of it. Oil prices plummeted in the 1980s, and the majority of Africa's new holes were abandoned and plugged and declared "commercially unviable." This is the industry's way of saying that there isn't enough oil in a particular field, or that what is there is too heavy or sour, to justify the expense of building an elaborate drilling and transportation infrastructure to bring it to market.

What makes today's African oil boom different from its less-successful predecessor is a combination of technological advances, high global demand, and consistently elevated prices. Traditionally, when prices are high and profit margins fat, oil companies have had the luxury of investing in research and development, coming up with technological innovations that enable them to drill for oil in ways that might never have been imagined a few years earlier. Just when everyone thinks the world is running out of oil and prices shoot up, the industry announces that it has invented a drill that can drill sideways, or one that can go down to 7,000 feet below sea level or, to take a recent example, that it has perfected better and cheaper ways to refine heavy Canadian oil sands and make them "commercially viable." It is a business strategy informed by a crude, almost Malthusian sense of optimism about resource management, but by and large, it seems to have worked. And in the early 1990s, as the first Gulf War pushed up oil prices, the big innovation to come on stream was deepwater drilling technology.

Offshore exploration is nothing new to the international oil industry. For much of the 1960s and 1970s, drilling platforms could be seen going up in the shallow tidal waters along California, Louisiana, and Texas, as well as in the North Sea, Brazil, and parts of western Africa. Any drilling that took place in depths greater than 1,000 feet was considered "deepwater" and depended on what was then cutting-edge technology. However, in the early 1980s, oil companies operating in the Gulf of Mexico began experimenting with new technologies that allowed them to drill to depths of up to 5,000 feet (four times the height of the Empire State Building). These technologies made possible the drilling of so-called subsea wells at the bottom of the seabed, miles from the shallow waters of the upper continental shelf. These wells would be serviced not by pipelines running to shore, but by a new animal in the world of offshore drilling: the floating production storage and offloading vessel, or FPSO. FPSOs are giant shiplike hulls that contain floating factories the size of several football fields where crude oil extracted from deep waters is brought for processing and production, stored in containers that can hold in excess of 2 million barrels, and then offloaded onto supertankers for transportation to refineries anywhere in the world. From start to finish, no one ever really needs to go onshore.

Along with improvements in deepwater drilling, FPSOs have made it possible for oil companies to venture 200 miles or more off the world's coastlines, into one of the last frontiers of oil exploration, the deep blue sea. So profound has been the impact that two-thirds of the world's new oil and gas discoveries now come from deepwater reserves. Drilling depths of 7,000 to 8,000 feet have already been achieved, and Chevron has probed for oil at a record depth of 10,010 feet (picture the altitude of the typical passenger jet about ten minutes before landing) in its Toledo prospect in the Gulf of Mexico. This deepwater and "ultra-deepwater" revolution has caused great excitement in the oil industry, with the supermajors in particular seeing a lifeline in the face of dwindling global reserves. In 2004 the highly respected Scottish oil consultants Wood Mackenzie and Fugro Robertson went so far as to predict that deep and ultra-deep reservoirs hold 181 billion barrels of undiscovered reserves—or more than twice the amount of oil discovered to date around the world.

One of the most striking features of this deepwater boom, moreover, is its geographic distribution. Although Asian and Pacific Rim deepwater is expected to make a bigger impact in coming years, at least 75 percent of the world's deepwater reserves are believed to be in the Atlantic basin—in effect, the waters off of Brazil, the Gulf of Mexico, and the West African "margin," the 5,000-mile coastline stretching from Senegal to Namibia. Together, these three areas of prolific deepwater activity are referred to in the industry as the "Golden Triangle," and account for 90 percent of the capital that oil companies have invested in deepwater exploration and production.

If you are a geologist, you will see no coincidence in these locations turning out to be the goose that laid the golden egg for the deepwater drilling industry. As long ago as 1930, German scientist Alfred Wegener looked at a map of the world, noticed that the eastern shoreline of South America and the western shoreline of Africa fitted together like pieces of a jigsaw puzzle, and suggested that the two continents had been joined in the distant past. In the decades since Wegener put forward his controversial hypothesis, geologists have come to recognize that South America and Africa were once part of the so-called "supercontinent" of Gondwanaland (or Pangaea), and that they became separated during the late Jurassic and early Cretaceous period, by a process of continental drift and oceanic rupture that left massive deposits of salt and sedimentary rock—along with such organic matter as seashells and fossils—along the two new continents' outer shelves. It is this organic debris that turned into hydrocarbons in what is now the deep waters of Brazil, the Gulf of Guinea, and the Gulf of Mexico.

Brazilian and Mexican waters remain under the tight control of their countries' national oil companies, Petrobras and Pemex, and the U.S. Gulf saw most of its big deepwater advances take place in the 1980s, but what the industry (slightly inaccurately) refers to as "West Africa"* has been a much more recent arrival to the party. The first really massive deepwater fields to be discovered in the continent's western margin were Shell's Bonga field in Nigerian Block OPL 212 and Total's Girassol field in Angola's offshore Block 17—both in 1996. Since then, dozens of fields have been discovered in depths of more than 1,000 feet, some of them going deeper than 7,000 feet.

One of the factors that held back development of West African deepwater in the early years was the lack of economic incentive for oil companies. In the late 1990s the price of crude oil was hovering around the historic low of $15 a barrel. Fresh off the assembly line, an

---

*Traditionally, West Africa is defined as the southern half of the bulge of land in the northwest, from Senegal to Nigeria and as far north as Mali, but not including the western reaches of North Africa, such as Morocco and Algeria. When the oil industry speaks of "West Africa," it is using the term more literally, to mean simply the western coast of Africa.

FPSO can cost $800 million, and operating a single deepwater well can easily cost $400,000 a day—sums that make even the biggest of the world's supermajors think twice before starting production on a new deepwater field during a time of low prices. But in the first years of the twenty-first century, all that has changed. Fears about future demand from countries like China, as well as continuing volatility in the Middle East, Venezuela, and Russia, have pushed oil prices to sustained, near-record highs of $50 to $80 per barrel, and the industry has proven far more willing to sink big money into a very promising region. Africa has finally become "commercially viable."

And the subsequent acceleration in the rate of international oil majors' investment in the continent's offshore fields has been breathtaking. In the year 2000, the industry spent $626 million developing deepwater fields in Africa. By 2003 that figure had risen to $4.5 billion and, according to British energy analysts Infield Systems Ltd., it is expected to top $6 billion in 2006—representing nearly half the global total for deepwater capital expenditure. During this period, ExxonMobil alone spent $3 billion just to develop its Kizomba A project offshore Angola. The number of FPSOs parked along the African coast has gone from none a few years ago to six today and will probably keep climbing. Dozens of new deep or ultra-deep fields are scheduled for development in Africa between 2005 and 2010.

Where the oil companies see big profits in Africa's offshore boom, many industry analysts also see a more strategic reason for excitement. They claim that by focusing oil exploration and production on deepwater reserves, all the difficulties present in the Niger Delta can be circumvented. Offshore, there are no angry villagers demanding compensation for oil spills or disrupted livelihoods. There are no ethnic militias kidnapping foreign oil workers, or gangs of unemployed youth stealing crude oil. And, perhaps most importantly, there are no international NGOs or shareholder activists demanding to know the role oil companies are playing in exacerbating conflicts between government forces and local communities. Far from either hostile natives or pesky Western pressure groups, a man can finally be left alone to drill in peace. Out in the deep blue sea, it is always just going to be you, your rig, and the silvery fishes.

◄◄·►►

Small wonder, then, that what was going on beneath the waters of the Gulf of Guinea began to attract the attention of a number of people in Washington without any obvious connection to the oil industry. As the twentieth century drew to a close, a growing number of lobbyists and lawmakers began suggesting that it might be time for the United States to take a close look at Africa, for possibly the first time in U.S. history. The reaction, as one of these lobbyists told me, was predictable. "You couldn't go knocking on doors in Washington wanting to talk about Africa. It was a lower priority than Antarctica." But then, in 2001, Texas businessman George W. Bush came to national power, and brought with him a vice president with many years of knowledge and experience in the oil industry and, many have suggested, a vested interest in its success. Suddenly there were a few more doors to knock on.

Almost immediately after Bush was inaugurated in January 2001, he made it clear that attaining energy security was a top priority of his presidency. In the months leading up to his inauguration, many parts of the United States had suffered severe oil and natural-gas shortages, and a series of dramatic electrical blackouts had struck some of the most heavily populated regions of California. Furthermore, for the first time in history, more than half of America's oil supply was coming from overseas.

One of Bush's first actions was to set up a task force called the National Energy Policy Development Group, headed by Vice President Dick Cheney, and often informally referred to since then as "the vice president's task force on energy policy," or simply "the Task Force." The job of the Task Force was to address what the Bush administration believed was the security "crisis" America would face in coming decades, and come up with a long-term strategy.

Perhaps ominously, the Task Force met in secret, and to this day the question of who took part is the source of acrimonious debate in Washington, with oil-company executives strenuously denying their presence in the meetings. In May 2001 the Task Force issued a report, which the White House released to the public under the authoritative-sounding name "National Energy Policy." The document contained a series of recommendations, the most controversial being that the United States open Alaska's pristine Arctic National Wildlife Refuge to oil exploration. One of its less-publicized conclusions, however, was that "West Africa is expected to be one of the fastest-growing sources of oil and gas for the American market."

A few months later, "the world changed"—or at least the world as Americans knew it. As the United States struggled to come to terms with the catastrophic terrorist attacks on its soil, the Bush administration told Americans to expect a long, complicated, and unconventional war against al-Qa'ida and its sympathizers around the world. Before long, neoconservatives within the administration began considering ways to prepare the American public to accept the idea that there might be a link between the September 11 hijackers and Iraqi president Saddam Hussein. In the midst of all this, in January 2002, a now-infamous breakfast symposium was held at the University Club in downtown Washington by a lobbying outfit calling itself the African Oil Policy Initiative Group (AOPIG).

In attendance at the AOPIG meeting were a number of officials from the Pentagon and the State Department, oil-industry executives and lobbyists, and a handful of diplomatic officials from relevant African countries. The keynote address was given by Representative Ed Royce, chair of the House Subcommittee on Africa, who stressed that "post-9/11, it's occurred to all of us that our traditional sources of oil are not as secure as we once thought they were," and added, for good measure, that it was "very difficult to imagine a Saddam Hussein in Africa." Walter Kansteiner, at the time the Assistant Secretary of State for African Affairs, the State Department's highest-ranking official for Africa, arrived late but gave his blessing to the festivities, saying it was "undeniable" that African oil "has become a national strategic interest for us." Shortly thereafter, AOPIG released a supporting document called "African Oil: A Priority for US National Security and African Development."

The timing of the AOPIG event, as well as the political agenda of its organizers, has given rise to speculation considering subsequent events in the Middle East. AOPIG and its activities were the brainchild of Paul Michael Wihbey, of the Institute for Advanced Strategic and Political Studies (IASPS). IASPS, which now appears to be mostly defunct, was a conservative Jerusalem-based think tank that also received heavy funding in the United States from the archconservative Bradley and Scaife foundations and the Hoover Institution. IASPS's close links to Israel's right-wing Likud party have led many to question whether AOPIG was merely a propaganda exercise, part of a coordinated effort to convince Americans that the United States no longer needed to be dependent for its oil on troublesome Arabs, and therefore no longer had to make concessions to Arab grievances in the Middle East.

IASPS makes little effort to hide its contempt for Arab dominance of the world's oil supply, or even for Arabs in general. Its Web site states explicitly, "West African oil is what can help stabilize the Middle East." Even more than that, it can "end Muslim terror" and deliver "a measure of energy security." Diversifying the country's energy dependency has been a goal of successive American administrations since the 1970s, and is hardly a radical policy. In the years since September 11, 2001, in fact, there has been a growing mood among Americans that the country is overly dependent on the Middle East for its energy needs. To a great extent, this reveals a fairly simplistic understanding of world affairs, exemplified by the incantation that the war in Iraq was really "all about oil," and the assumption that if only the United States didn't rely on Middle East crude, all its problems in the region would somehow magically disappear. Simplistic or not, however, it is a point of view that has rapidly gained traction and given rise to what some have referred to as a "geo-green" movement in the United States—a renewed interest in conservation and alternative technologies, such as ethanol and hydrogen, not so much for the sake of saving the planet, but for the sake of energy independence and disentangling the United States from questionable Middle East alliances. In that sense, the IASPS agenda was very much in tune with the American zeitgeist in the early years of the twenty-first century.

Still, a quick trawl through the IASPS Web site makes for some troubling reading. An impassioned mission statement puts forth the need to oppose strenuously those policies that are "undermining the elementary truths of human order." The institute places its analyses in terms of stark moral choices, reserving its most animated criticism for "Western elites" who it claims have "sought bases of agreement with Nazis, Communists and now with terrorist Islam." The mission of the institute is to "expose the deception" of these elites, who have set up "contrary deities" in the form of "Democracy, the Open Society, Equality, [and] Freedom." In 2004 IASPS organized a seminar on "The Convergence of Western Elites and Islam," at which adjunct scholar D. Y. Anaximander denounced "the Hollywood crowd" and "the professors" for their "solidarity with the Muslims." Anaximander railed against "arch-Hate Master Noam Chomsky . . . the Jewish anti-Semite leader of the liberal Jewish legions within the Western Elites." The choice of upper-case nouns was his own.

The institute's American board, meanwhile, reveals a great deal about its historical and intellectual roots. It includes such geriatric Cold Warriors as William Van Cleave, who in the 1970s was a member of both the Committee on the Present Danger and the notorious "Team B," a hawkish group of government officials that overrode CIA and State Department objections in an effort to exaggerate the

Soviet menace. Sound familiar? Team B's other members included such young Turks as Paul Wolfowitz and Donald Rumsfeld.

However, IASPS's chief claim to fame among students of the neoconservative movement is not its paranoia or ideological purism or wacky political and racial ideas, but its association with a certain document. In 1996 IASPS published a series of recommendations for the then-incoming prime minister of Israel, Binyamin Netanyahu, under the title "A Clean Break: A New Strategy for Securing the Realm." The position paper, which advocated that Israel adopt a more aggressive posture toward the Palestinians and its Arab neighbors, was the product of a study group led by none other than Richard Perle, who went on to become chair of the Pentagon's Defense Policy Board and one of the architects of the war in Iraq. Other signatories to the document included such American neocon shining lights and future members of the Bush administration as Douglas Feith and David Wurmser. "Clean Break" contained a number of chillingly prophetic recommendations, such as "containing and even rolling back Syria" and "removing Saddam Hussein from power—an important Israeli strategic objective in its own right." As a result, the document has taken on some mystique among those who study the neoconservative movement, as a supposed blueprint for the Pentagon hawks, a kind of Federalist Papers for the people who planned and pushed for the Iraq war in 2003—many of whom, the implication goes, displayed a greater sense of allegiance to Israel than they did to the United States.

The exact merits of this argument are the subject of ongoing debate in Washington, much of it dyspeptic and radioactive, and mercifully beyond the scope of this book. What is certain, and far more relevant here, is that much of what AOPIG and Paul Michael Wihbey have written is characterized by the kind of breathless ideological purity that has come to be seen as a hallmark of the neoconservatives— and that Wihbey, during 2002 and 2003, was able to drum up significant media coverage through AOPIG.

Many commentators in the United States now look at those two years as a historic low point for American media, a time when even such venerable newspapers as the *New York Times* allowed themselves to get swept up by the march to war and published stories clearly fed to them by administration hawks. Wihbey's agenda, which included the idea of building a U.S. military base in the Gulf of Guinea, was no exception to this phenomenon. As Americans were nudged into believing the 9/11 hijackers were connected to a soon-to-be-nuclear Saddam Hussein, a slew of articles in major U.S. newspapers spoke in excited terms about how African oil might soon "replace the Middle East" as a source of U.S. energy security, casually parroting the mantra that the Gulf of Guinea was "the new Persian Gulf." Perhaps the most priceless example of this was a lengthy *New Yorker* article published in October 2002, under the headline OUR NEW BEST FRIEND: WHO NEEDS SAUDI ARABIA WHEN YOU HAVE SÃO TOMÉ? The country in question, a pair of tiny volcanic islands in the Gulf of Guinea home to barely 150,000 people, had yet to drill a single barrel of oil.

Wihbey's other major contribution was to bring together people with very few interests in common, help them catch the Africa bug, turn them into petro-evangelists, and then set them loose to preach the gospel to the uncoverted. So, on a January morning in 2002, Democratic congressman William Jefferson, a member of the Congressional Black Caucus, found himself making common cause with Walter Kansteiner, an evangelical Christian conservative and old Angola hand, who had made a name for himself in the 1980s by condemning Nelson Mandela and the ANC as violent Marxists.

In recent years, IASPS appears to have wound down its activities, and many of the more mainstream figures who lent their support to AOPIG have migrated to less ideologically driven channels. The more establishmentarian, less neoconservative strain of African petro-evangelism that has emerged in recent years is epitomized by the Center for Strategic and International Studies, an influential bipartisan think tank whose many high-profile members include four former secretaries of state. In July 2003, CSIS put together what it called a "Task Force on Rising US Energy Stakes in Africa," cochaired by David Goldwyn, who had served as Assistant Secretary of Energy for International Affairs in the Clinton administration. The Task Force's job was to assess whether an argument could be made for a higher level of U.S. engagement with Africa's booming oil producers.

Perhaps not surprisingly, the answer was yes. In its first report in March 2004, the CSIS Task Force described Nigeria, Angola, and other energy-rich African nations as being "at a promising moment of opportunity," and suggested that, with "enhanced, high-level US engagement," this moment could be "successfully exploited." Its follow-up report in July 2005, titled "A Strategic US Approach to Governance and Security in the Gulf of Guinea," recommended making security and governance in the Gulf of Guinea an "explicit priority" in U.S. foreign policy, and projecting a "robust, comprehensive policy approach to the region."

CSIS advocated uniting officials from the relevant U.S. agencies, such as the Departments of State, Defense, and Energy, under a single approach. The effort should be driven by a new "special assistant" to the president and secretary of state, modeled on the existing special adviser for Caspian Basin energy, who could be responsible for organizing such initiatives as an annual African energy summit, and whose presence would send a message about the priority Washington placed on the region. Essentially, CSIS asked for an African oil czar.

◄◄•►►

Even the most fervent African oil evangelist, however perfervid his desire to see America reduce its dependence on the Middle East, knows in his heart that Africa is no silver bullet and that the deepwater revolution is not everything it is cracked up to be.

Though deepwater drilling allows companies to circumvent the potential for a violent operating environment, it does not allow consumer nations to wash their hands of the complex web of problems that oil wealth can bring to a developing African nation. The apparently serene and predictable operating environment of deep-sea drilling masks a far more troubling threat to the long-term stability of oil-exporting countries in Africa than a few kids with guns. The threat is one with no immediate impact on oil-company operations, but one that, if not guarded against, can thoroughly eviscerate the social and economic fabric of a country.

The "curse of oil" is a term that has become hugely fashionable in recent years among those concerned about poverty reduction and the effect of resource booms on developing countries. Those who have put forward this idea of the "paradox of plenty" have argued that it can take many forms, from the exacerbation of preexisting armed conflict to the encouragement of corruption to the neglect of traditional industries and agriculture. (So wide-ranging and inclusive is the concept, in fact, that skeptics accuse it of being nebulous and crude.) Certainly, in Nigeria, oil wealth has brought endemic conflict. But, for a growing number of activists and NGOs, the real "curse" of oil is not political or military instability, but economic degradation.

That oil wealth could be a curse seems counterintuitive. When an oil bonanza is discovered in a struggling African country, the instinctive assumption is that it can only be a good thing; that it will result in a rapid improvement in the lives of the people; that suddenly there will be money for hospitals and vaccines and schools and roads; and, even more than that, everyone will be rich. To the contrary, however, studies suggest that real GDP and the population's standard of living nearly always decline where oil is discovered. Between 1970 and 1993, for example, countries without oil saw their economies grow four times faster than those of countries with oil.

To understand how this paradox operates, we must journey back to the decades following World War II, when an explosion in the global demand for oil began delivering petrodollars to the treasuries of otherwise obscure and underdeveloped countries that just happened to be blessed with large hydrocarbon deposits. From Mexico City to Baghdad to Caracas, government planners rubbed their hands together with glee and dreamed of the possibilities. Great cities were built or rebuilt from scratch, with towering office blocks and modern universities. But as time went on, it became increasingly obvious that the much-touted oil wealth was bringing only economic stagnation and cycles of debt to the world's oil-exporting countries. By the early 1970s, the pattern had become so clear that the Venezuelan oil minister, Juan Pablo Pérez Alfonzo, who had been one of the original architects of OPEC, described the oil gushing from his country's reservoirs as "the Devil's excrement."

Part of the explanation for why countries flush with oil cash were not experiencing the rapid growth they had expected can be found in a phenomenon economists call the "Dutch disease." Originally used by the *Economist* in 1977 to describe the collapse of the manufacturing sector in the Netherlands following the discovery of natural gas there in the 1960s, the term now refers more generically to the negative effects of exchange-rate appreciation on an economy that becomes suddenly overreliant on one type of export commodity (usually an extracted natural resource).

What does this mean in plain English?

When a developing country suddenly finds itself selling a highly valuable natural commodity (such as oil) on the international market, the money it receives from buyers will not come in the form of Nigerian naira or Angolan kwanzas, but in dollars and euros and pounds, and so the country quickly finds itself flooded with foreign currency. This glut of foreign exchange artificially inflates the value of the country's own currency, which means that suddenly imported products become much cheaper and everyone rushes out to buy foreign goods, which are perceived (usually accurately) to be of better quality than domestic products. Special K and Weetabix replace millet and boiled cassava as dietary staples, imported steaks and Chivas Regal become more fashionable than the local curried goat and national-brand beer, and a burgeoning elite begins splashing out on Land Cruisers and portable MP3 players.

It appears the country has become rich overnight, but cassava farmers and goat herders find fewer people will buy their products. The natural reaction of these agriculturalists is to abandon their failing rural livelihoods and flock to the cities, where they have heard there is big money to be made. Once there, however, they end up selling cigarette lighters and sticks of gum on the street or, if they are lucky, driving taxicabs. Meanwhile, this mass urban migration devastates the country's traditional farms and small cottage industries. And, in what is probably the bitterest irony of all, thanks to the collapse of the agricultural sector, life in the big cities becomes increasingly reliant on expensive foreign food, which is largely out of reach to these new arrivals from the hinterland, who find themselves dependent on government handouts and international food aid. In short, given a sudden infusion of foreign currency, a country that was once a regional breadbasket and net exporter of food can quickly turn into one that is unable to feed itself.

This might almost be tolerated as an acceptable price to pay for the overall increase in the country's wealth. Unfortunately, oil is a finite resource and only when it runs out is the true cost of the Dutch disease felt. Almost as quickly as they arrived, the dollars and euros disappear from the nation's economy, the national currency rapidly depreciates, and the power of consumers to purchase foreign goods collapses. Worst of all, as there is no longer a traditional agricultural or light-industrial base for the economy to fall back on, the country finds itself in much worse shape than it was in before oil was discovered. In extreme cases, the process leads to a total dependence on foreign aid.

To some extent, the evidence of Dutch disease can be seen in Nigeria, where between 1965 and 1975 petroleum production went from accounting for just 5 percent of government revenue to a whopping 80 percent, bringing with it a dangerous dependence on oil-price volatility, the bottoming out of traditional farming and manufacturing industries, and the inauguration of a never-ending cycle of external debt. However, in Nigeria the violence and political unrest that have accompanied oil exploration have been so extreme as to overshadow this economic fallout. For a textbook case of the disease, one so iconic and so faithful to the prescribed pattern that it might as well have been created by economists in a test tube, one need travel no more than a hundred miles from the Niger Delta, to the Republic of Gabon. Just on the other side of the rich blue waters of the Gulf of Guinea, decades of petro-prosperity have left a country totally unprepared for what happens when the oil runs out, as it is now beginning to do. There, when talk turns to the "curse of oil," the story is not one of AK-47s and kidnapped Filipinos and jerry cans loaded with illicit crude under cover of night. In Gabon, the curse of oil has nothing to do with guns and speedboats and ultimatums, and everything to do with the price of Brie.

◄◄-►►

If you come to Gabon directly from Nigeria, you might be forgiven for thinking you have landed in an African version of the French Riviera. The road into the capital, Libreville, from the airport quickly takes on the august form of a grand boulevard snaking along the oceanfront, complete with lampposts and a grassy median decorated with the flags of neighboring countries. On one side as you head south into the city center are such smart, triumphal buildings as the presidential palace, the Hotel Intercontinental, and a handful of discreet embassy buildings. On the other side is nothing but the rhythmic lapping of shin-high waves and the muggy angel-hair embrace of a tropical breeze.

Gone is the sight of legless cripples, crawling on their bare hands through lanes of traffic like teams of crazed, foreshortened gymnasts, competing for prizes of loose change. Gone, too, is the smoke billowing from mountains of trash that have gone uncollected so long that residents have set fire to them. And gone completely is the shouting and the jostling and the barely suppressed rage that seems to flow through Nigeria's streets like a howling flume of molten lava from morning to night. In its place is a distinctly languid holiday feel and an unmistakable air of genteel French provincialism left over from colonial times.

This outward splash of easy prosperity has much to do with Gabon's small population and sizeable oil reserves. In a country that is only a little smaller than Nigeria and pumps 265,000 barrels of oil a day, there are not 130 million people to share the oil wealth, but just over *one* million, making Gabon's per capita income of $6,500 one of the highest in Africa. (Compare it with Nigeria's $678.)

As if to reinforce the point, in the chic downtown of Libreville, there are white people everywhere. When the French colonial authorities left this part of Africa in 1960, thousands of functionaries and technocrats and oil-company employees stayed behind, happy to take advantage of the new government's palpable lack of hostility toward them. Today the French community is 10,000-strong in Libreville (a city of barely half a million people), helping to ensure that a portion of the posh Nombakélé district along the water remains a timeless re-creation of Nîmes or Avignon, with young women in summer frocks and designer sunglasses darting in and out of pastry shops.

My first duty, like any African traveler exhausted by the flies and the heat and the rusty Kalashnikovs of weeks on the road, was to myself. Within an hour of arriving in Libreville, I found myself, blinking like a newborn deer, in the air-conditioned aisles of the Score supermarket. Score is an old and shabby chain by Western standards, barely competitive in France in the age of the *hypermarché*. The Libreville incarnation would hardly pass for a corner deli in the United States today. But as far as I was concerned, it could have been the Galéries Lafayette.

Eyes bigger than my stomach, I surveyed the piles of refrigerated Brie and Camembert and Port Salut, the neat rows of jars sealed tight

to protect the foie gras and goose-liver pâté inside, the shelves of Petit
Écolier chocolate biscuits and Kinder eggs, and the separate counter
where fresh baguettes and pastries were sold. In the fruit and vege-
table section, there was no shortage of the produce integral to a European diet: tomatoes and cucumbers, parsley and broccoli, apples
and oranges, and even several varieties of squash. What seemed conspicuously absent, however, were tropical fruit and vegetables—the
yams, mangoes, and pineapples. Looking around, I couldn't even see
any bananas. Perhaps they were sold out? Perhaps in another section?

No, sorry, came the smile and the shake of the head when I asked.
"Pas des bananes." No bananas.

Well, no matter. Perhaps bananas aren't the kind of thing one buys
at Score. Perhaps expats come here for their Lindt and Yoplait but buy
their local produce at local markets. I ventured back outside, into the
tomblike humidity of the February afternoon, to look for one of the
women carrying large plates of bananas on their heads who seem
ubiquitous in southern Africa. But here, too, I was out of luck. The
only street vendors in sight were the lumpen young men from Mali or
Niger or the Congo who make up much of Gabon's unskilled workforce, and all they had on offer were fake Rolexes and snakeskin belts.

I soon learned that it is extremely difficult to buy a banana in
Gabon. During a week and a half in the country, I feasted on the most
delicately prepared beef bourguignon and rack of lamb, always served
with haricots verts or gratinée potatoes. But never did I manage to
find a bunch of bananas for sale.

This is not to say there are no bananas in Gabon. Quite the contrary—there are literally millions. About half the country's tiny population lives in Libreville, while most of the rest reside in three or four
towns that dot the jungle interior. Where the population centers stop,
there is nothing but miles and miles of misty virgin rain-forest, inhabited by gorillas and lizards and thick with banana trees. All one has to
do is walk to the edge of town, into where the bush begins, and there

is no shortage of banana trees to choose from. The country is bursting with sweet, soft bananas.

So what happens to them? The ones that are not picked by chimpanzees and baboons turn yellow, then brown, then black, and then fall to the ground and fertilize the jungle floor. Before oil was discovered in Gabon, the country was self-sufficient in bananas. By 1981 it had become almost totally dependent on bananas imported from neighboring Cameroon. One could not ask for a more vivid image to associate with the Dutch disease than that of a vast jungle nation where there is no one available to pick bananas off the trees.

And it's not just the bananas. In the early 1980s, when Gabon was at the height of its oil production, and globally it seemed that cheap oil was a thing of the past, the country imported a staggering 96 percent of its food. Even eggs were flown in, as no one in this cash-flooded nation could be bothered to raise chickens. Today Gabon is dependent on imports for 60 percent of its food needs—still an uncomfortably high figure in a country where the oil is beginning to run out.

On paper, as well as on the streets of Nombakélé, Gabon does a fairly convincing impression of being a rich country. With a daily output of 265,000 barrels of oil, Gabon is the fifth-largest producer in sub-Saharan Africa, earning between $2 billion and $3 billion a year from oil—ostensibly a huge sum of money for its small population. And everywhere along the waterfront is the opulence that oil wealth has brought. Smart boutiques selling everything from toys to women's accessories to the latest electronics line well-paved streets; Lebanese restaurants serve patrons sitting under umbrellas at Parisian-style outdoor tables; and an evening at one of Libreville's trendy nightclubs can easily set you back several hundred dollars. Year after year, Libreville is ranked as one of the world's four or five most expensive cities, consistently pricier than London and New York.

But both the healthy macroeconomic indicators and the lifestyle along the Libreville waterfront are representative of a prosperity that is only skin-deep. Venture inland a little, past the absurd Ozymandian edifices of various government ministries, and you are quickly back in "Africa," hurtling down dusty, boulder-strewn streets, through teeming shantytowns marked by open sewers, mangy goats, and corrugated zinc roofs. Scratch away at the impressive per-capita income of Gabon, and you will quickly see a country whose faulty economic stewardship and overdependence on oil has left its citizens signally unprepared for life *après petrole*. Unemployment in Gabon is a whopping 40 percent, and the United Nations says two-thirds of the population lives on less than $1 a day.

That decline is on its way now, and coming faster than anyone is ready for it. In 1997 Gabon hit what will almost certainly turn out to have been its peak, at 371,000 barrels of oil a day, and was the third-biggest producer in sub-Saharan Africa, behind only Nigeria and Angola. The country's output has declined significantly since then, and in 2005 stood at a mere 233,000 barrels per day. Today Gabon has been overtaken by both Equatorial Guinea and Sudan and is in a dead heat with Congo and Chad for the rank of fifth-largest sub-Saharan producer.

◄◄·►►

But Dutch disease is only part of the explanation for why Gabon and other resource-rich countries have failed to turn their cash wealth into meaningful development for their citizens.

The first economist to tackle seriously the question of why oil-exporting countries did not have the fastest-growing economies in the world was Hossein Mahdavy in 1970. Responding to the economic crisis in his native Iran, which at the time was swimming in oil revenue but stagnating under the burdens of slow growth and external debt, Mahdavy suggested that countries that are reliant on oil exports for the majority of their income be described as "*rentier* states," which he defined as "countries that receive on a regular basis substantial amounts of economic rent."

For at least a century, the term *rentier* (pronounced RON-tee-yay) had been used to describe people (generally elites) whose income came not from manual or professional work, or from entrepreneurship, but from collecting rent on property they already happened to own. Such people were known collectively as a *"rentier* class," a phrase that carried a slightly negative connotation in the years after the Industrial Revolution, conjuring images of a lazy, landed class of yesterday's men living on inherited wealth and showing little inclination to cultivate skills, develop industry, or engage in economically productive activity. Mahdavy was the first to apply the term to an entire nation, suggesting that a country that sat back and collected income from the oil that foreign companies drilled out from under its soil was one whose primary function on the world stage was not that of a laborer, farmer, skilled artisan, or entrepreneur, but rather that of a wealthy landlord.

In the years that followed, other economists elaborated Mahdavy's definition of the *rentier* state, and helped clarify why such a state was prone to economic weakness and stagnation. For a country to be described as a true *rentier,* it was argued, not only did external rents have to become the dominant source of income, but these rents would have to be generated by the activities of only a few people, and accrue almost entirely to the government directly. For example, tourism—an activity that requires the participation of many people, who benefit directly by their participation—could never be the basis of a *rentier* state, though it involves the collection of external rent. The obvious impact of a *rentier* arrangement is that it divorces the government and its management of the economy from the day-to-day needs and the economic activity of the population.

Under normal circumstances, the economic productivity of citizens is extremely important for a government because the government is dependent on domestic taxation for its revenue. In other words, politicians have a direct interest in encouraging industry and generating wealth in the country because wealthier people means more tax revenue for the treasury, which means more money invested in public services that win popularity for elected officials, and in programs to encourage more agricultural and industrial productivity. In a *rentier* economy, however, the situation is reversed, so that the state is no longer reliant on the economic productivity of its citizens for its revenue, but itself becomes the main *source* of revenue in the domestic economy. Under political pressure to spread the newfound wealth, politicians take the path of least resistance, which is not to get involved in complicated schemes to promote the country's traditional agricultural or industrial base, but rather to create lots of new government jobs for people. Politicians believe that by doing this they are avoiding the temptation simply to hand out cash, giving citizens something for nothing, instead creating a sense of ownership in the country's future on the part of the new government employees. In fact, they are creating a bloated bureaucracy, staffed by people who are paid to do very little besides push papers and introduce obstacles and inefficiencies into the functioning of the country's already-faltering economy.

In such a shift, from a "production state," in which productivity and growth are prized by the government as building blocks of a healthy tax base, to an "allocation state," in which the government functions as a vast gravy train of handouts and pet projects, elected politicians are seen as having access to vast sums of money and must spend much of their days fending off requests for help and favors from extended family members. In a country in which the government is reliant on a tax base, citizens view corruption and cronyism as an affront and are more likely to organize against it. In a *rentier* state, by contrast, corruption is seen by many as the only way to get ahead and avoid missing out on their "piece of the action." The population largely tolerates the situation because it sees the state as a sugar daddy, a source of "free money," rather than a publicly accountable body.

Even if a *rentier* state is led by inspired and visionary politicians determined to behave with integrity and the best interests of the population at heart—always a big if—there may be a real limit to how much they can achieve. To begin with, an extreme dependency on oil exports for national revenue leaves a country vulnerable to the frequent, unpredictable fluctuations in the price of oil. In a country like the Republic of Congo, for example, where petroleum accounts for a staggering 90 percent of export revenue, it is difficult to plan budgets from one year to the next or to put long-term macroeconomic strategies in place when you know that, within a couple of years, oil can dive from $80 a barrel to $20 a barrel. But even if a politician takes the brave decision to invest an oil windfall in developing traditional agriculture and industry, it is not always clear how he can go about doing so. If you are one of the world's poorest nations and your traditional economic mainstay is primitive cassava farming, such a small cottage industry does not have the capacity to absorb a sudden investment of billions of dollars. Therefore, even the best-intentioned leaders in a *rentier* state find that at least a portion of the country's windfall is better invested in overseas brokerage accounts, where a reasonable return on investment seems guaranteed over the long term. And so there the money sits, collecting interest and gathering dust, dipped into occasionally by unscrupulous politicians or spent on military hardware to put down rebellions, but rarely used to develop cassava farming.

As devastating as the effects of the *rentier* state are on the economy of a resource-rich nation, they pale in comparison with the political effects. When leaders no longer feel the need to tax their citizens to raise revenue, they become far less interested in what those citizens think about them, and unresponsive to complaints about their job performance. Meanwhile, citizens who pay little or nothing in taxes become far less interested in politics, and begin to see the cash-rich state as simply a source of lucrative contracts and easy favors. Before long, it appears that hardly anyone has a stake in how the country is governed, and a dangerous democratic deficit can open up.

And even when oil price fluctuations or the Dutch disease leave governments unexpectedly strapped for cash, politicians in a *rentier* state are reluctant to raise money by means of taxation. Perception on the street is that the country is awash in oil money and that if all this wealth is not enough to run the country, then clearly a lot of it has been disappearing into the pockets of politicians. Faced with an electorate that is highly resistant to being taxed, and painfully aware that it would face defeat in any free and fair election following its mismanagement of the economy, the government of a *rentier* state often falls back on authoritarianism and paramilitary force as the only way to retain power and maintain law and order. Oil states often therefore become highly militarized. From 1984 to 1994, for example, the military expenditures of OPEC countries as a percentage of their total budgets were three times that of developed countries, and up to ten times that of developing countries that were not members of OPEC.

None of this, however—not the exposure to oil price fluctuations, not the Dutch disease, not even the breakdown of democracy—is as corrosive or as threatening to the fabric of a nation as the subtle psychological effect that a *rentier* economy can have on the population. Economists have defined this *"rentier* mentality" as a shift in the general attitude toward work and compensation such that, in the words of the economists Hazem Beblawi and Giacomo Luciani, "reward becomes a windfall, an isolated fact." In other words, people who witness vast amounts of money appear out of nowhere and be shared unfairly begin to view personal wealth as the result of an *accident,* or an ability to be in the right place at the right time, rather than as a *reward* for hard work.

This severing of the perceived link between work and remuneration can have a deeply destructive impact on the economy. Small businesses like corner shops and small factories are abandoned as their proprietors try their hands at real estate or other speculative quick-buck ventures in an attempt to strike it rich off the coattails of the booming oil sector. Young, well-educated citizens give up on pursuing professions or starting businesses and instead seek out petty government patronage in the form of lucrative ministerial appointments. Manual labor is considered demeaning and is left to immigrants, who simply send what they earn to their families back home. And—perhaps most destabilizing of all—everyone is filled with a nagging suspicion that it is somebody *else* who is making the *really* big bucks, a feeling that if the country is so rich in oil, then *everyone* should be driving BMWs. In the poorest and most deprived communities, there develops a profound resentment when this does not happen. And at all levels of society, from top to bottom, there sets in a cynical, short-term, grab-what-you-can mentality that can take generations to shake off.

Gabon's recent history is best understood as that of a small, struggling country that won the lottery, went on a binge, and is just now waking up to an almighty hangover. From 1966 to 1976, Gabon's oil production jumped almost tenfold, from 29,000 barrels a day to 226,000 barrels a day, and all at a time when Middle East politics and rising global demand were helping to push oil prices through the roof. The 1973 Arab oil embargo was a particular fillip to Gabon, which saw its state budget nearly triple from 1974 to 1975. At one point, oil was providing as much as 90 percent of public revenue and there seemed no end to the good times. Even in the early 1980s, when Gabon's production was beginning to decline, higher-than-ever oil prices kept the party going, and encouraged a sort of hideous collective *nouvelle richesse*. By 1984 Gabon had become the world's leading per capita consumer of champagne. Eight thousand magnums were wheeled out just to celebrate the wedding of a prominent politician's daughter. At a now-legendary four-day summit of African heads of state that Gabon hosted in 1977, the president, Omar Bongo, dropped an eye-popping sum of $800 million (75 percent of the country's national budget for that year) building fifty-two villas for his guests and flying in a fleet of Rolls-Royce limousines and armor-plated Cadillacs for the occasion.

But when it comes to examples of lavish and gratuitous spending, nothing rivals the Trans-Gabonais Railway. In 1972 President Bongo began construction of this 400-mile railway from Libreville into the country's jungle interior, declaring that in Gabon, "We want to do in a few decades what others took centuries to do." In other words, Gabon was going to show the world that it could fast-track into the space of a few years the kind of industrialization and development that European nations had achieved over the course of many generations. Economists cautioned that the $1.4 billion railway was going to be a reckless waste of money, a vanity project that would cripple Gabon's chances at real development. But Bongo was adamant. "The Trans-Gabonais will be built," he insisted when foreign lenders refused to back the project. "It will be built by one means or another, with the help of one country or another."

In the end, the Trans-Gabonais took fourteen years to build and cost $4 billion, triple the original estimate. Six million trees were felled, fifty bridges built, and four thousand workers (only half of them Gabonese) employed in the construction. Almost from the day it opened for business, the Trans-Gabonais proved unprofitable and required a $60 million-a-year subsidy just to stay up and running. Bongo had built his longed-for railway but, as virtually everyone had predicted, it had plunged the country into a vicious cycle of debt from which it has yet to emerge, proving only that "development" is not something that can be delivered overnight, and that Bongo's dream of an African Rome would certainly not be built in a day. In the words of one Gabon analyst, Bongo's spectacular mismanagement of his country's oil boom has left it with "little more than a handful of rusting factories, a choo-choo train, and a massive government debt."

Bongo may have a poor head for development economics, but there is no denying that he is one of the shrewdest and most skilled politicians in Africa, if not the world. Using a combination of strategic alliances with foreign leaders and an impressive ability to buy the loyalty of potential opposition figures at home, Bongo has managed to remain Gabon's president for nearly four decades, without resorting to brutality or violence. Catapulted into the presidency in 1967,

Bongo immediately created a one-party state and won elections in 1973 and 1986 with 99.59 percent and 99.97 percent of the vote. Though the international community has since pressured him into re-instituting multiparty democracy, he has continued to win elections handily, the most recent, in December 2005, giving him yet another seven-year mandate at the age of seventy. With the death of Togo's Eyadema Gnassingbe in January of that year, Bongo even inherited the mantle of *"le doyen d'Afrique"* (Dean of Africa), the nickname given to the continent's longest-serving head of state—an impressive achievement in a land of Qadhafis and Mobutus. If he survives his current term (and there is no reason to believe he won't), Africa's 5'1″ Energizer Bunny will have been president of his country for forty-five years.

A canny political survivalist, Bongo has cultivated close ties with key foreign leaders, helping to add to his mystique and stature at home and convince the average Gabonese citizen that no one else is fit to be president. After all, what other head of state anywhere in the world can claim to have been personally received by Mao Tse-tung, Chou En-lai, Deng Xiaoping, and Jiang Zemin? Is anyone else in this small jungle nation really capable of stepping into those shoes?

However, what Bongo knows as well as everyone else in Gabon is that he owes the largest part of his longevity in office to the protection of the French, who gave him his political start in life and who, through successive changes of government of their own, have helped him stay in power. A member of the tiny Bateke tribe, Bongo was born in a village deep in the jungle interior and became a petty clerk in the colonial postal service in the 1950s. Only when he joined the French military in 1958 did he become close to influential Gaullists and important members of the so-called "Foccart network," a group of spies and ex-soldiers that then managed France's Africa portfolio. Like most of the native elites who eventually took power when French Equatorial Africa was decolonized, Bongo was groomed from early on by the French, educated and assimilated into the system as part of an effort to avoid producing a generation of radical anti-French nationalists during the liberation process. The strategy worked perfectly in Gabon, whose independence was negotiated in a nonviolent, orderly process that began with the Brazzaville Conference in 1944 and culminated in 1960 with the solemn lowering of the French tricolor over Libreville. The entire liberation process was, in the words of Gabon scholar Alan Yates, "an elite undertaking," one that "deposited freedom, as it were, in the laps of the Gabonese elite."

Throughout the 1970s, France cemented its dominance over an independent Gabon. The country was ruled with an iron fist by an oligarchy that consisted of Bongo, his right-hand man Georges Rawiri, French ambassador Maurice Robert, head of Elf-Gabon Maurice Delauney, a contingent of French mercenaries trained by Pierre Debizet, and the notorious Foccart network. Members of the *reseau Foccart* even provided French and Moroccan soldiers to command the Presidential Guard, a 1,500-strong force drawn mostly from Bateke kinsmen loyal to Bongo.

In recent years, Bongo has attempted to outmanuever the French by diversifying his country's commercial and political links. In 2004 alone, for example, Bongo welcomed Chinese president Hu Jintao, Morocco's King Mohammed, and Brazilian president Luiz Inácio Lula da Silva, and was himself invited to the White House to meet George W. Bush. Relations with the United States have quietly grown closer, which has irked the French; but relations with China have also improved, which has irked the Americans. For its part, France has spent much of the last decade reassessing its relationship with Africa, showing signs of turning inward and focusing its diplomatic efforts on the European Union. Increasingly, France appears prepared to forgo its traditionally paternalistic approach to its former colonies and abandon the vanity that speaking for Africa will help the French state retain a semblance of credibility and relevance on the world stage. Even more important, outgoing President Jacques Chirac may be the last of an old breed of French politicians with close personal ties to Africa's leaders.

For now, though, France continues to dominate Gabon's political, economic, and military affairs. Thirty-eight percent of Gabon's exports go to France and 61 percent of imports come from France. As is the case in nearly all of Francophone Africa, Gabon's currency is pegged directly to the French franc (and, by extension now, the euro) at a fixed exchange rate. A steady flow of "development aid"—much of it tied to lucrative procurement contracts with French companies—comes into the country through the French agency for overseas cooperation. And, in a robust gesture of friendship and solidarity, Paris keeps a regiment of eight hundred Marine infantry troops stationed permanently outside Libreville. Officially, they are there to provide technical assistance and cooperation; unofficially, they help consolidate the air of global importance and invincibility Bongo has tried to establish over the years. In 1990, for example, when civil unrest broke out in the oil town of Port-Gentil, the French Marines were deployed quickly—ostensibly to evacuate French nationals working in the oil sector and to protect French oil-company installations, though the message sent by the show of force was not lost on anybody.

Gabon, however, it should be emphasized, is by no means a police state or a savage dictatorship. Bongo's forty-year tenure as president is not the product of torture chambers and gulags and secret police files. Bribery, favoritism, and crude coalition-building are the preferred methods of coercing political opponents into seeing things the government's way. One Gabonese sociologist told me, "Every time ordinary people try to create a political party, the powers that be find out who their leader is, call him in, and hand him a fat envelope. He quickly shuts his mouth and you never hear from them again." As a result, the Gabonese public sees itself as led by an entrenched and self-serving political class, but recognizes there is no credible alternative to rally behind. The few times that moderately successful opposition parties have formed, they have been dismissed by a skeptical population as vanity projects of former Bongo loyalists who have fallen out with the government over the exact price of their allegiance.

During presidential elections, opposition parties in Gabon rarely articulate platforms that say much more than that it is "time for a change."

Though venality rather than thuggery is the dominant characteristic of the Gabonese political system, to an outsider the distinction is not always clear and the country can look like a classic tin-pot dictatorship—a sort of banana republic (with the part of the bananas played by Cameroonian understudies). In Libreville, dissent seems to come only in whispers, and the presence of Omar Bongo is felt everywhere. Rare is the hotel lobby, restaurant, bank, or hospital without a portrait of the young, elfin Bongo grinning down magisterially over visitors. Even the four-lane boulevard that runs along the coast, one of the city's most important arteries, isn't immune from the theatrics of *Bongoisme*. On a weekday morning, I sat roasting with four other passengers in an airless taxi, as rush-hour traffic stopped for twenty minutes so that the president's helicopter could make its landing on the lawn of the presidential palace.

One evening, at my hotel, after I had dined under a portrait of the great man, I placed a call to Nicaise Moloumbi, the head of Croissance Saine Environnement, which describes itself as an independent advocacy group promoting healthy and sustainable development. As soon as I began to introduce myself, Moloumbi interrupted to ask where I was staying. He seemed relieved when I told him I was at the Hotel Monts de Cristal. "That's the best place to stay!" he exclaimed, in a tone that suggested a joke I wasn't quite getting. He told me to meet him in half an hour at his office, which, bizarrely, he said I would find on the second floor of my hotel.

Baffled, I went to the second floor and, sure enough, found an unmarked office where Moloumbi waited, looking nervous. Most NGOs I had visited in Africa operated out of dingy, sweltering concrete offices with shuttered windows and disintegrated sofas for guests to sit on. Moloumbi's office, however, was air-conditioned, comfortably appointed, and covered in pictures of Omar Bongo. Here was Bongo with Mitterrand, there was Bongo with Chirac, and everywhere was Bongo at official state functions. I had assumed that, as an NGO director, Moloumbi would have something critical to say about the government. Instead, he wasted no time telling me how the French had ruined Gabon, and complaining of the voraciousness of Elf (now Total)—a standard position for supporters of the government. When I steered the conversation to the responsibility of the Gabonese government, Moloumbi had only glowing words for Bongo, and his attempt to move the country away from French influence. "Many people in Gabon don't realize that he's giving them the opportunity of a lifetime."

The hotel, I would later learn, was owned by one of the president's sons, and really not the place to engage in frank discussion about the government's performance. When I requested a meeting with Jean-Silvio Koumba, one of Bongo's top advisers, he met me in the lobby and immediately suggested that if I wanted to talk about oil politics, I should come around to his house, where we could speak more freely.

We met on the veranda of his home in a quiet suburb of Libreville on a muggy Sunday morning and his wife quickly brought out a tray of cold grapefruit sodas. Koumba is an economist who, once upon a time, had written a doctoral dissertation at the Sorbonne on the political economy of petroleum in Gabon. I began by asking him what he thought oil had done for Gabon over the past fifty years.

He took a long time to answer. "Honestly?" Pause. "Nothing." It was true, he admitted, that the economy had remained stunted and uneven over time, but he was unwilling to acknowledge the government's role in perpetuating the Dutch disease. "In the 1970s, we spent money on useless projects, like bringing in cattle from other parts of Africa that couldn't survive in Gabon—a French initiative."

I brought up the Trans-Gabonais, thinking the subject might be a prickly one, but Koumba saw the railway as a towering achievement of Bongo's presidency. The World Bank and the international community had wanted a comprehensive road network built instead,

Koumba noted, and if Gabon had borrowed the money from France to do that, all the construction contracts would have gone straight to Paris. "This was a wise decision for which we thank the president today. If we had built the roads like they told us to, the French would have taken everything."

It seemed an utterly surreal justification, based on nothing more than an intransigent bloody-mindedness about anything French. Gabon, one of the richest countries in Africa, still lacks a road network to connect its major cities. The only real option for traveling between Libreville, Port-Gentil, and Lambaréné, where Albert Schweitzer had his hospital, is to fly over the dense forest canopy in a commercial airplane. Only Libreville and Franceville are connected by the Trans-Gabonais. It hardly takes an economics doctorate to understand why the lack of such infrastructure might hold back a country's economic development, and yet so fragile is national pride, so deep and institutionalized resentment of the French, that even this inexcusable failure is spun as a victory for sovereignty and self-reliance.

Perhaps Koumba was on to something. Perhaps after centuries of slave raiding, commercial exploitation, and colonial domination, the ability of an African government to make its own mistakes—no matter how costly to the population—is worth far more to the collective psyche than the sight of a freshly asphalted four-lane highway cutting through the jungle, built by white people. Perhaps. Koumba was surely correct, though, when he told me, "Many Gabonese can't turn their backs on Bongo because they know he's the only one who knows how to deal with the French."

I would have been remiss in my duty as the Western naïf, however, if I hadn't asked Koumba the obligatory question about democracy before I left. After thirty-eight years with the same president, was it not time for a changing of the guard? Or at least a more genuine multiparty democracy? Koumba stared at me icily and tapped his pen. "Democracy?" he said finally. "Since 1990 [when multiparty democracy was established] we haven't built a single kilometer of road.

Everything has to go through the *Collège* and the Parliament now, and all day long it's 'we want the road here, we don't want a road there' and so on." He was just getting warmed up. "You know what 'opposition' means in Gabon? You didn't name the son of a certain man from a certain village to a particular government post, and the village got angry. There's your opposition." Koumba motioned to his children, who were buzzing about on the veranda. "Surely you know by now that democracy doesn't exist in Africa. If I tell my son to go away and stop bothering us while we're talking, he does exactly that. He doesn't see it as an opportunity for debate. He doesn't form an opposition party. If you read the Bible, you see that leaders are anointed by God Himself." The conversation ended soon after that, and Koumba led me out of his house, past a large faded wedding photo of Omar Bongo that hung over the dining-room table.

One of the few places I did hear a bit of open dissent was the dusty hillside campus of Libreville University, a short taxi ride from the gaudy glamour of Nombakélé. Officially, the university is called Omar Bongo University, and to get there, you drive down Omar Bongo Triumphal Boulevard, past a collection of bizarre-looking official buildings (the twenty-story Ministry of Environment and Natural Resources is shaped like a tree) and the new parliament building being erected with the help of a generous grant from the Chinese.

On campus, I met Pierre-Fidèle Nze Nguema, a respected sociologist and author of an important social history of Gabon. Nguema was as critical of France and of Elf as everyone else I had met in Gabon, but he was not willing to let his own government off the hook for failing to prepare for the day when the oil runs out. "People went crazy" during the oil boom of the 1970s and early 1980s, he said, "but we did nothing to take advantage of the oil money. They just wanted to exploit it as long as it was there. What we didn't do was to create industries. We never developed other kinds of companies. Even Nigeria did that much. Even they developed companies that today do something other than exploit oil." *Even Nigeria.* Nguema had dropped the N-bomb. In Africa, there is no way to deliver your verdict more explosively when assessing the impact of oil on your country than to draw an unflattering comparison with the experience of Nigeria.

But perhaps we should not be too quick to pass judgment on Bongo's stewardship without some recognition of how difficult it can be to diversify a *rentier* economy. As one senior French official I spoke to in Libreville put it:

> It's not so easy just to replace oil revenue. Manganese is barely 3 percent [of the economy]. Timber employs a lot of people but in fiscal terms, it's not much. Agriculture? The Gabonese are not really farmers, they are foresters. Tourism? The country has a lot going for it, but there are some important constraints. Flights are expensive. There's the question of visas, the quality of hotels. At the moment it's just for the rich. Mass tourism isn't going to come here tomorrow. Services? No spirit of entrepreneurship. Big business is in the hands of the French and small and medium enterprises are in the hands of the West Africans and the Lebanese.

It seemed a sober assessment. In fact, the international community may need to come to terms with the idea that Gabon probably represents a realistic "as good as it gets" for the impact of oil on African development. The "curse of oil" has been almost entirely an economic one for Gabon, having to do mostly with a lack of preparation for the post-oil era. The country has experienced no violence, no ugly tribal conflicts, no real brutality in its forty-six-year history as an independent nation. Even corruption is more of the petty kind than on the order of the billions missing in nearby Angola, Nigeria, or Equatorial Guinea. Stories about Gabon in the international press are virtually nonexistent and, sadly, in a continent of plagues and famines, wars and genocide, this splendid anonymity, this tacit acknowledgment that nothing much ever happens here, is probably as close as Africa gets to a real success story.

Other oil-rich African countries, including some of Gabon's immediate neighbors, have not been so fortunate.

◄◄·◆·►►

When I touched down in Brazzaville, capital of the Republic of Congo (Gabon's neighbor to the south and also known as Congo-Brazzaville), I was pleasantly surprised. The streets were freshly painted, and flanked by neat rows of posts to keep children and stray goats from wandering into traffic. It looked oddly prosperous for a country that had fought two brutal civil wars in the last ten years and is generally described as one of Africa's most forgotten tragedies (upstaged in the international press by its far larger neighbor, the Democratic Republic of Congo, formerly known as Zaire and now sometimes as Congo-Kinshasa). Two days before I landed, Brazzaville had been the scene of a fresh outbreak of fighting between government soldiers and the fearsome "Ninja" rebels of Pastor Ntumi's Committee for National Resistance, so I had come prepared for a slightly more grisly welcome than I got.

I quickly learned, however, that Brazzaville's prosperity was purely cosmetic. French President Jacques Chirac had been in town a few weeks earlier, and the government had decided to spiff up the place— or at least the roads that ran between the airport and the key government ministries. Just out of sight of the Chirac route were buildings strafed with bullet holes and amputees hobbling along in silence, some crawling on the ground like amphibians for lack of crutches.

Save for the occasional street orphan coming up to beg for change, the streets of Brazzaville buzz with the eerie silence of a deserted financial district on a Sunday morning. Once-grand office towers stand empty and abandoned, their windows smashed out, and orderly offices long ago torched and looted. Along wide avenues, trees grow into and out of the remains of smart French-style boutiques, wrapping themselves around mangled iron girders. Crumbled porticoes serve as accidental tombstones for the decomposed corpses everyone knows lie just beneath the weeds.

Even the hustlers in Brazzaville carry out their duties with quiet resignation. Every few yards, someone will walk up to you and, without a word, show you a leather belt or a pair of plastic sunglasses, or point at your shoes in the hope that you will want them shined. From as much as half a mile away, a taxi driver will tap gently on his horn—just once—and slow down in anticipation at the sight of a white man who might need a lift. Scores of taxis circle aimlessly around Brazzaville's streets, but they are the only traffic in sight, and they are always empty. Many drivers, not wanting to waste fuel, stand on street corners, washing and shining their green and white cabs, which are easily the cleanest in Africa.

Brazzaville is one of the poorest and most miserable of Africa's capital cities, a place where electricity is sporadic and running water a luxury few residents can afford. In a city of 800,000, only 60,000 are registered customers of the state water utility, but even they regularly go days without service, forced to pay taxi fares on top of their bills to go searching for clean water. As for the taxi drivers themselves, they begin their mornings before the sun has come up, parked for hours in front of filling stations, waiting for fuel, which is always in short supply. Primary-school attendance, which was over 90 percent before the wars, has slipped to 44 percent. In December 2004 the government began the slow process of catching up on the salaries and pensions of state employees, most of whom had not been paid for two years. Malaria and infectious disease are rampant, and street children have become a real social problem. In 2003 a global survey declared Brazzaville the world's worst city to live in, behind even Baghdad.

And towering above it all—above the corpses and the bullet holes, the shelled-out buildings and the zombielike taxis, above even the cathedral on the hill—is Elf Tower, the only office block that wasn't damaged during the war. Here, as every Congolese knows, is where the country is more or less run. Because to describe the Republic of Congo as an enclave economy is a little like calling the Pope Catholic. In most years, oil accounts for 70 percent of the country's income, 80 percent of the state's annual budget, and some 90 to 95 percent of Congo's export revenue. What's even more remarkable, however, is that some 70 percent of Congo's oil production is conducted by the French multinational Total (formerly Elf). With stats like that in mind, it's not hard to understand why the place made such an effort to look nice for Chirac.

It's also not hard to understand why Congo-Brazzaville, like Gabon, has fallen prey to the ravages of the *rentier* mentality. Successive governments have proven more interested in prestige projects and such profligate white-elephant schemes as airports, gymnasiums, and radio stations than in roads and schools and hospitals. The current president, Denis Sassou-Nguesso, has been particularly fond of the trappings of sovereignty, playing host to such events as the African athletics championships or a festival of African film, and opening costly embassies in foreign capitals, all designed to show the world that Congo punches above its weight. Agriculture, which still employs 40 percent of the population, goes largely ignored by a government preoccupied by the lucrative rent-seeking opportunities associated with the offshore oil industry. As a result, only 2 percent of Congo's arable land is farmed. "What's going to happen after the oil is gone?" asked opposition leader and one-time presidential candidate Joseph Kia Mboungou, when I dropped by his modest office tucked behind the Libyan embassy. "It is suicide. We are going to commit collective suicide."

◄◄·►►

In a way, though, oil has already been the death of Congo. *Içi, le petrole tue,* goes a popular saying in Brazzaville—"The oil here is lethal." It's a reference to the country's disastrous civil wars in the

1990s and early 2000s, which killed 10,000 people and remain the cause of deep traumatization among the civilian population. In Congo, it is taken for granted that access to oil revenue was at the root of the fighting. Former Elf executive Loïk Le Floch-Prigent even said as much in an extended interview he gave in 2001. "Arms were delivered," he told the French journalist Éric Decouty. "People died. Month after month, as their oil is being sold, the Congolese see a part of their money go directly to Elf to pay for these arms. This shameful business went on for four years, and did anyone care?" The World Bank agreed, stating in 2000, in its more-diplomatic language, that "management of the country's rich natural resources" was a primary factor that had fueled the conflict.

Among those who have won control of Congo's natural resources, there has been very little bashfulness about indulging in the perquisites of power. In 2005 the French investigative newspaper *Le Canard Enchaîné* revealed that President Sassou-Nguesso's nephew Willy had bought a $3 million apartment in Paris and finished it with expensive rugs, plasma-screen TVs, and several tons of imported marble. The twelve-bedroom, nine-bathroom pied-à-terre boasted seven kitchens and a garage filled with Jaguars, Porsches, and an Aston Martin DB9—accoutrements obviously appropriate to the young Sassou-Nguesso's role as head of the Congolese National Maritime Transport Agency, a Total subcontractor. The president himelf, meanwhile, caused an uproar later that year when he ran up a $295,000 hotel bill in one week in New York. Records showed that Sassou-Nguesso, a former Marxist, paid $8,500 a night for his suite at the Palace Hotel and racked up $12,000 in room-service charges. His fifty-person entourage included his butler, his personal photographer, and his wife's hairdresser. The purpose of the visit: Sassou-Nguesso was to make a fifteen-minute speech to the United Nations General Assembly.

For Pastor Ntumi and the Ninjas, who remain loyal to deposed prime minister Bernard Kolélas, such reports can only be salt in the wound. The rebels continue to camp in the jungle on the outskirts of

Brazzaville, in the impoverished region of Pool, and rely on smuggling and banditry to make ends meet. Despite a March 2003 peace accord, sporadic fighting between government forces and Ninjas means the region has never had a chance to recover from 2002 when, during the worst of the fighting, more than 100,000 residents of Pool (a stunning 99.8 percent of the population) fled their homes. Entire villages have been destroyed, harvests have been disrupted, and only 8 percent of households collect water from a tap. The United Nations has described the situation in Pool as a "forgotten crisis."

Brazzaville and the coastal oil city of Pointe-Noire, Congo's political and economic capitals and clearly its two most important cities, are both served by frequent flights from Paris, but between them lies an unpaved road that years of warfare have left in the most appalling condition (the entire national road network, in fact, is only 5 percent paved). As if to make matters worse, in 2004 and 2005 the Ninjas began attacking passenger trains as they clattered along the decrepit rail line that runs between the two cities. The Congo-Océan Railway, built in 1934, was one of the great achievements of French Equatorial Africa, but today it is an unreliable, denuded shell of its former self. One hundred of its 311 miles pass through Pool, and frequent breakdowns, floods, and hijackings have turned one of the world's classic train trips into a grueling, unlit, twenty-hour nighttime journey through the rain forest.

I had dearly wanted to take the train, and had even met the eccentric Frenchman who heads the railway, Jacky Trimardeau, to seek his security advice. Trimardeau, a burly, sweat-drenched beast of a man, had offered to make bodyguards and plainclothes police officers available to me, as well as undercover Ninjas, but in the end, the once-weekly trains didn't fit my schedule and I had to plump, like most expats, for the forty-minute, $320 flight aboard Trans Air Congo.

Arriving in Pointe-Noire, it is hard to believe you are in one of the global oil industry's top destinations. The airport is a corrugated tin shack, its "baggage claim" an aggressive, sweaty mosh pit of trolleys and limbs run by an angry gendarme who occasionally throws over-enthusiastic passengers against the wall when they try to climb over the counter to grab at their bags. I had to beat back several teenage boys to get hold of my suitcase before one of them could "help" me with it, and in the process was nearly knocked over by a giant bunch of unripened plantains that was flying over the counter.

Once in town, though, Pointe-Noire becomes extremely familiar. Behind the grassy train depot, where the Congo-Océan comes in, there is a rubbish-strewn beach, fringed with palm trees and crawling with people. On the horizon, a line of gas flares shoots from the off-shore rigs, and on the beach itself is a gated compound belonging to Total. In town a long, wide avenue—the avenue Charles de Gaulle—sweeps past most of the hotels and travel agents, almost American in its lack of bends and curves or unexpected treasures. At every corner are the usual suspects—joints with names like Royal Flush and Fortune's Club, Safari Grill and Carnaval—each sporting gaudy Christmas lights and mirrored glass doors, and promising all the sleazy delights your patriarchal soul can handle.

They are there mostly to cater to the "oilfield trash," a subculture of fat, sweaty, mostly Anglo-Saxon men I was beginning to see at every stop now, and the Africans who think their cultivation is something to aspire to. The "trash," who wear the nickname as a badge of pride, are generally men in their thirties and early forties, all bushy mustaches and baseball caps, patterned, open-necked shirts and denims, bulging guts and cheap cigarettes. They have found the promised land, where they can make enough money to walk around with nineteen-year-old Naomi Campbell look-alikes on their arms. Where there is oil—whether it's Gabon or Galveston, Congo or Caracas—there are the trash. And where there are the trash, there are the bottom-feeding locals—the prostitutes and the tear-jerking cripples, the fake-Rolex vendors and the shoeshine artists, the hustlers and the beggars and the whores. And then there are the Lebanese.

And everyone wants a piece of the action.

There is another side to Pointe-Noire, though, even if it is not always visible between the rigs and the flares and the baseball hats. In
the years since the last civil war ended, a tentative civil-society movement has sprung up and has begun calling attention to what it considers a culture of corruption and a lack of transparency in the
management of the country's oil wealth. In 2002 the Catholic Church
in Congo wrote an open letter to President Sassou-Nguesso complaining that "the Congolese people do not know much about how
much our country receives from this black gold, and even less about
how the revenues are managed. What it does know is the price of oil
is measured not in barrels or dollars, but in suffering, misery, successive wars, blood, displacement of people, exile, unemployment, late
payment of salaries, non-payment of pensions." In 2004 the church
was bolstered by the British NGO Global Witness, which documented in detail a pattern of kickbacks, influence peddling, and questionable accounting in Elf's business activities in Congo-Brazzaville.*

By and large, however, activism in Congo has been held back by
a combination of fear and intimidation, stymied by the population's
traumatization. "In places where people have had to listen to the
sound of cannons being fired, it is true, you will find people who say
that it's better on balance to just carry on with things as they are than
to risk the fragile peace we have achieved," said Joseph Mandzoun-
gou, a former treasury minister who now runs a youth-training center in Pointe-Noire. "People who have been displaced from their
homes, naturally, have very little appetite for any of this. But I will tell
you that is exactly what the powers that be are counting on."

Again and again in Pointe-Noire and in Brazzaville, church leaders and activists told me how difficult it was to begin openly discussing corruption and oil-revenue transparency. "Here, for a long
time, people thought that if you touched the question of oil, you were

---

*"Time for Transparency: Coming Clean on Oil, Mining and Gas Revenues" (*Global Witness*, March
2004), pp. 21–39.

going to die," said Roger Bouka Owoko, a prominent human-rights activist in Brazzaville. "That is, until September 2002, when the bishops became the first to criticize openly the management of oil revenue. That was a profoundly important day in our history. It demystified the whole issue of oil management. Until then it had been a sacred cow, a taboo." Even then, the bishops had hesitated. Their letter had been ready in June, but it took them three months to work up the courage to publish it. "They felt under a lot of pressure," says Owoko. "They came to me and said, 'Can we really do this?'"

"It was hard at first," the archbishop of Brazzaville, Anatole Milandou, confessed to me. "The politicians didn't like it. They said it wasn't the role of the church, and so on." Milandou was a soft-spoken man who gave the impression of someone deeply reluctant to become involved in politics, but he told me that in 2002 he felt it had become impossible to ignore reality any longer. "We knew years of war here, and everyone knows that the focal point of all that war was oil. Fundamentally, people fought over control of oil. And it is the prophetic duty of the church to intervene when people are suffering. So we intervened."

As for Denis Sassou-Nguesso, he has proven far too sophisticated to make martyrs out of the transparency campaigners, preferring instead to play the anticolonial card so beloved of African despots. Asked by the French magazine *Jeune Afrique* in 2005 about his country's reputation for high-level corruption and the lack of transparency in the oil industry, Sassou-Nguesso replied archly that "as far as I know, the Enron scandal didn't take place in Brazzaville." Asked about the lack of vigorous opposition in Congo, Sassou-Nguesso was equally smug. "You think it's a good thing to insult one's leader on every street corner? You think it fitting that the little local magistrate can come knocking on the door of the president to summon him to court? In Africa, we don't like that sort of thing, we don't admire that. Our values are that we respect our leaders. In this regard, Africa will never imitate Europe, you can be sure of that."

African leaders long ago learned that these kinds of glib answers
can silence Western critics full of postcolonial guilt and fear and buy
a politician a little breathing room with internal critics. Occasionally,
though, this rhetoric means something. As I was about to find out,
there is at least one country in Africa where this kind of amour propre was something of a national obsession.

CHAPTER 3

"A COUNTRY IN AFRICA"

ANGOOOOOOOOOOOOOOOOOOOLA!

If the blue and white Toyota Hiace minivan had come any closer, I might have gone down in history as the first person to be run over by a speeding vowel.

SOMOS ANGOLAAAANOS! SOMOS ANGOLAAAANOS!

Like all the blue-and-white Toyota Hiace minivans clattering around Luanda, this one was on its last legs. The sliding door had been slid back and forth so many times that it no longer worked and was now held open with lashings of duct tape. Both headlight casings had been replaced with squares of white tissue paper. All four wheels looked as though they might fly off at any moment. And its rusty exhaust pipe trailed a giant cloud of chocolate-colored smoke.

Had this been an ordinary day, up to fourteen passengers would have been shoved inside this standard form of public transport, looking weary and uncomfortable in the afternoon heat. Now, in addition to the usual passenger load, a group of teenage boys, beer bottles in hand, were trying hard not to fall off the roof as the van swerved around corners. Several more hung on to the van's empty window frames, dangling into the path of oncoming traffic. And on the back bumper two more boys stood holding a giant Angolan flag so that the large yellow Marxist logo flapped in the breeze. The driver kept one hand against the horn, helping to turn the threadbare van into a boisterous victory party on wheels. It was October 8, 2005, and the Angolan national soccer team, the Palancas Negras, had just defeated

Rwanda away in Kigali, 1–0, to become one of five African nations to qualify for a berth at the World Cup. Something approaching total pandemonium was breaking out on the streets of the capital.

Boys on 125cc motor scooters darted between minivans and pickup trucks piled full of drunken fans, breaking into forty-five-degree wheelies in every patch of clear road. In the large roundabout circling the cathedral, they accessorized their high-speed wheelies with impressive acrobatics, putting both feet up on the handlebars, or striking graceful synchronized swimming-style poses, with left arm and right leg raised at ninety-degree angles.

In the space of a few short hours, everyone in Luanda had discovered their inner Evel Knievel, and it was taking ample reserves of agility and blade-sharp reflexes to avoid being run over. It was hard to look around, though, without getting at least a small lump in the throat. After all, until 2002, Angola's citizens had, with the exception of a few brief moments of calm, been at one another's throats for nearly four decades—whether in the long and divisive battle for independence from Portugal, or the epic civil war that broke out even before the colonial powers had left in 1975. For twenty-seven years Angolans had been forced to take sides between the ruling MPLA, founded as a Marxist liberation movement and backed by the Soviet Union and Cuba, or the rebel UNITA, backed at various times by the United States and apartheid-era South Africa. And, in one of Africa's heartbreaking coincidences, forty is the average life expectancy in Angola. Thus, in a country where more than 500,000 people have been butchered by their fellow citizens and 4 million more forced to flee their homes and villages and livestock, today only the very oldest can remember what life before war was like.

A mere three and a half years after the civil war came to an end, people were dancing in the streets and shouting that they were all Angolans now. *Somos Angolanos*. National television showed footage of the 370 lucky fans the government had flown to Kigali to watch the match. Spectators waving MPLA and UNITA flags could be seen sitting side by side in the stands. You would have had to have a heart of tar not to get just a little choked up.

It is not always easy for Americans to understand how much victory on the soccer pitch means to people in the rest of the world. Our own sporting obsessions tend to be purely domestic, though we declare the winners "World Champions," and during international sporting events like the Olympics, there is often a palpable lack of interest among the American public. But in virtually every other country on the planet, the ability of eleven men to put the right number of balls in the back of a net is a historic national achievement—one that says, "Yes, we've had our share of problems and we may not be perfect, but the world counts us out at its peril."

Perhaps nowhere was this more true in 2005 than in Angola. Despite years of bloodshed, despite the very name "Angola" having served for decades as shorthand for the death and destructiveness of civil war in Africa, Angolans have remained among the proudest of Africans. Boasting one of the continent's biggest and (sadly) most experienced armies, but also blessed with prodigious reserves of oil, diamonds, gold, timber and copper, Angola has long given its people reason to believe it has the potential to be one of Africa's real powerhouses, on the level of Nigeria, South Africa, and Egypt. "Angola is a country in Africa, not an African country" is an expression you quickly get used to hearing from Angolans and one it would be a mistake to dismiss as a piece of fussy amour propre. One senior Western diplomat, who had spent nearly twenty-five years in Africa before being posted to Angola, told me he found the self-confidence in Luanda a "big surprise."

So when the final whistle blew in Kigali, and the streets of Luanda erupted in celebration, it was not with the blinking disbelief of an African backwater that had somehow slipped through the cracks, but rather an almost-indignant sense of vindication. The match's result was a reminder of what everyone had known all along—that it really *had* just been the war holding the country back all these years. It was about bloody time; at last—at long, long last—the world was beginning to see what Angola was really all about.

But the truth is that the world had already begun to see what Angola was all about, and it had nothing to do with soccer or the platitudes of national reconciliation. For much of the past decade, Angola has played host to one of the most feverish oil booms Africa has ever seen, and one driven almost entirely by deepwater discoveries. In 1985 Angola was producing 232,000 barrels a day of crude, making it a midrange player in Africa. By the late 1990s, as deepwater drilling brought new offshore fields onstream, that figure had jumped to 750,000, where it stayed until 2001. But by the end of 2005, thanks largely to gigantic deepwater fields brought into production by ExxonMobil and Total, Angola was churning out 1.3 million barrels a day, nearly double its output four years earlier. Even the most conservative estimates now put daily production at 2 million barrels by the end of 2007, moving Angola within striking distance of Nigeria for the distinction of being Africa's largest oil producer. In the early 1990s, Angola's total oil reserves were estimated to be 3 or 4 billion barrels. Today, analysts talk of 15 billion barrels, barely a drop of it onshore.

Angola is now a darling of the international oil industry, and Luanda has been transformed into one of the hottest business destinations in the world. Flights into the city, whether from Johannesburg or Lisbon or London, are consistently overpriced and overbooked, and even the most expensive hotel rooms are virtually impossible to get. Booking my accommodation six weeks in advance, I barely secured a room at the Hotel Vice Rei, where I paid $110 a night for live roaches, sporadic running water (generally brown in hue), and paint that dangled off the walls in large strips. When I complained to other expats, I was told to count my blessings—a room that cheap in Luanda usually doubled as a brothel. A massive luxury high-rise hotel had just opened and another was being built, but they were already booked up through the end of the year. So severe is the shortage that oil companies often book blocks of $250-a-night rooms two months in advance, on the off chance they might need to fly someone in.

With the price of oil breaking one record after another in 2004 and 2005 and the country's output levels rising almost as quickly, Angola has found itself flush with foreign cash. In 2005 oil generated 90 percent of all Angolan exports by monetary value, 80 percent of government income, and 50 percent of gross domestic product. Ten billion dollars flowed into the government's coffers from oil sales, up from $5.6 billion the year before. According to the government, this led to the first state budget surplus in the country's thirty-year history.

To get an idea of just how intense the interest from oil companies has been in Angola, one has only to look at the signature bonuses offered during a bidding round held in 2006 by the Angolan authorities. Traditionally, such signing bonuses are offered by oil companies as a symbolic gesture of their commitment and interest in a particular exploration license, a reflection of how keen they are to be awarded the acreage. Since the money is paid to the government before a single well is even drilled, and must be written off as a loss if the field proves uncommercial, companies generally exercise considerable caution in the amounts they offer up, with signing bonuses usually in the range of $10 million to, at the high end, $100 million. In this latest bidding round, however, oil firms offered up a combined total of some $3.1 billion between them—an astronomical figure that broke all records for the highest amount ever offered for exploration acreage anywhere in the world.

But it's not just sales of crude oil that have given rise to the giddy business atmosphere. With some of the biggest drilling projects still in the early development phase, there is a steady demand for oil-service companies and suppliers of peripheral goods and services. More important still, after decades of civil war, there is an enormous amount of work to be done on the country's infrastructure, with everything from roads to schools to sewage systems needing to be built from scratch. Seeing a country full of money and of needs, foreign contractors have been pouring in by the planeload. In 2004 *non*-oil and diamond investment in Angola topped $400 million, compared with $160 million the previous year.

Between the oil boom and the postwar reconstruction boom, Angola's economy has been experiencing growth spurts that are the stuff of dreams in Europe and North America, to say nothing of sub-Saharan Africa. The country's economy grew by a robust 15 percent in 2005, and another 14 percent in 2006. The forecast for 2007 is a whopping 31 percent, giving this war-ravaged "country in Africa" the slightly bizarre distinction of having the fastest-growing economy in the world.

All around Luanda, the outward signs of this little economic miracle are unmistakable. Luxury apartment buildings, complete with skyview penthouse units, are going up along the waterfront, and shiny Hummers accessorized to the teeth with DVD players, tinted windows, and bling-bling wheel trims speed down the city's rubble-strewn streets, scattering half-naked children out of the way. In fact, private-car ownership, virtually unheard-of during the first thirty years of the country's miserable history, has become the most sought-after status symbol in Angola. Luanda must have more driving schools per capita than any other city on earth; on any given day, half the vehicles on the road appear to be steered by student drivers with their instructors at their sides. And the souped-up SUVs cruising the streets with their subwoofers throbbing can sometimes give Luanda the feel of a main drag in New Jersey on a Saturday night.

With their imported Land Cruisers and mansions in foreign countries, the wealthiest Angolans, nicknamed the "100 families," indulge in some of the most extravagant lifestyles in the world. In 2003 an independent Angolan newspaper revealed that seven members of the presidential elite could each boast more than $100 million in assets, and estimated the personal wealth of President José Eduardo dos Santos as being "several hundreds of millions of dollars." This kind of money clearly makes dos Santos the richest man in Angola, but so extensive are his overseas assets that he is rumored to be the sixth-richest person in Brazil.

But the contrast between the lives of dos Santos and the 100 families on the one hand, and their neighbors in Luanda on the other, could not be more stark. During the civil war, some 4 million Angolans fled their homes, many seeking refuge in the capital. Luanda was built by the Portuguese to accommodate 400,000 people, but by the late 1990s, the population had swollen to more than 3 million. Today most of these displaced rural people live in unofficial squats and haphazard shantytowns known as *musseques,* with no leases or property titles or even personal-identity documents that can establish who they are. The United Nations estimates that 80 to 90 percent of Angola's urban residents live in homes with no clearly defined legal status. Needless to say, the city is overwhelmed, unable to provide them with basic services. Nearly half of Luanda's residents must buy water from private vendors—a staggering proportion even by African standards. A survey done in 1998 showed that the poorest quartile of Luandans spent 15 percent of their incomes on water, while the richest quartile spent only 3 percent.

In fact, the only real indication that Angola's *musseca*-dwellers and displaced rural families have that they are citizens of the world's fastest-growing economy is the extraordinary inflation with which they have had to contend in recent years. Throughout much of the mid-1990s the country endured increases in the consumer price index of over 1,000 percent a year. In 1995 inflation stood at 3,780 percent. The last time Luanda was included in a survey of the world's most expensive cities, in 1998, it came in at fourteenth place. With treble-digit inflation continuing through 2003, few doubt that Luanda is today far and away the world's most expensive city.

For foreign companies, who can hardly expect their employees to live in corrugated tin shacks or under bridges like the majority of Luandans, this can pose a real problem. Extreme housing shortages mean that in central Luanda, rent for a basic house with running water, electricity, and a phone line has hit $15,000 a month—enough to make New Yorkers and Londoners pipe down and realize how good they have it. But even more daunting than the price tag from the perspective of an oil company is just how hard it can be to find accommodation for employees. International companies have been known to pony up two or even three *years* of rent in advance to secure adequate premises.

Thus, a handful of ultrawealthy Angolan elite and an expatriate community whose bills are by and large picked up by their companies back home live on imported "luxuries" such as little $3 bottles of water and $2 pots of yogurt, while the rest of the population lacks the proverbial pot to piss in. In 2003 fully two-thirds of the country's citizens lived below the internationally recognized poverty line of $1.70 a day, and one in four struggled to survive on less than 76 cents a day—defined as "extreme poverty." Even something as rudimentary as a public taxi system, ubiquitous elsewhere in Africa, is nonexistent here. The only taxis available are a fleet of shiny, air-conditioned jeeps operated by Macon Taxi that charge an eye-popping $15 to drive barely a mile. I eventually found a good-natured driver who generously agreed to charge me only $120 a day for the privilege of being taken round in his banged-up old Toyota, but he never got tired of reminding me how lucky I was. The going rate for drivers in Luanda is $300–$400 a day.

And if the newfound wealth has failed to make its way to the city's overcrowded *musseques,* then it has been even less successful at reaching the hinterlands of this vast country, where the worst fighting of the final years of the war took place, and whole villages can only be reached by helicopter, thanks to blown-up bridges and roads littered with landmines. In a country of barely 12 million people, anywhere from 3 million to 8 million unexploded landmines are still buried—an obvious and tragic obstacle to the ability of rural communities and their livestock to survive on a day-to-day basis, not to mention the ability of humanitarian-relief workers to provide basic care to these communities. At the time of writing, there were believed to be villages unreached since the war's end, many of them so cut off by decades of fighting that they don't appear on official maps of the country.

Landmines and unexploded ordnance, however, are only the most tangible legacies of this brutal conflict. Officially, for example, one-quarter of Angolan children die before their fifth birthday (almost all from easily preventable diseases), but many experts believe even this grim statistic is an underestimate, since public-health officials have never been able to access the most isolated and vulnerable rural communities, where mortality rates are certain to be highest.

Whatever their limitations, however, certain statistics paint a powerful picture of how furiously and comprehensively the country's infrastructure, skilled labor base, and all-around ability to function were destroyed by the civil war. In a country that regards itself as one of Africa's up-and-coming powerhouses, only 16 percent of government employees have completed high school, for example. A country larger than Germany, with a population greater than that of metropolitan Los Angeles, has fewer than six hundred doctors. Every year, 20,000 Angolans die of malaria, the inevitable victims of a health-care system that just can't keep up. And the future looks even bleaker given that 45 percent of Angola's school-age children are not reached by the education system. In one of the world's wealthiest and most important oil-producing countries, barely half the children even have schools to go to.

Remarkably, though, as recently as 2001, the government's combined spending on health, education, water, and sanitation was a mere 9 percent of the national budget—the bulk of its money going to military expenditures or mysteriously unaccounted for. Even the International Monetary Fund, not known for its fondness for government spending on social sectors, expressed concern over this percentage. It has since begun to rise steadily, but remains far below the 30 percent typically seen in southern Africa.

What is of greater concern than the amount of state money allocated to social sectors, however, is the way in which these sums are spent. In the years between 1997 and 2001 scholarships for Angolan students to go abroad accounted for 18 percent of the education budget— more than was spent on technical and higher education combined. Clearly, foreign scholarships are not the best use of government education funds in the struggle against underdevelopment, as they are disproportionately handed out to the children of the elite, many of whom use their foreign university degrees to further their career prospects abroad rather than return home to help their countrymen out of poverty. Similarly, 13 percent of the country's health-care budget—nearly as much as the 17 percent spent on the nation's primary-care network—was spent on an expensive medical-evacuation service that allowed the 100 families to access state-of-the-art health care in Europe and North America. Angola's once-Marxist political leadership has proven so efficient and adept at turning the country's petroleum endowment into a source of personal enrichment that many Angolans now refer to them darkly as the "oil *nomenklatura*."

Between 1999 and 2004 the influential British campaign group Global Witness issued a series of reports detailing significant discrepancies between the oil revenues received by the Angolan state and the funds that made their way into the national budget, as well as a systematic pattern of official corruption, offshore money-laundering, illicit and highly profitable arms deals, and opaque agreements that mortgaged the country's future oil revenues to foreign banks. In the most recent of the reports, Global Witness drew on internal IMF documents to show that, for the years 1997 to 2001, $4.2 billion was left unaccounted for by the Angolan government's budget procedures.

To put this in perspective, it is worth remembering that Angola's gross domestic product during these years averaged $7 billion to

$8 billion a year, meaning that over a five-year period, an amount greater than half a year's national GDP disappeared. An appropriate analogy would be an American president admitting halfway through his second term in office that $6 *trillion* was missing from the U.S. Treasury, and then categorically refusing to publish any relevant documentation, or even to discuss the matter publicly.

For the Angolan government, the negative publicity generated by Global Witness and others represented the final straw in a fifteen-year battle of wills with the International Monetary Fund, which had been trying to nudge the Angolans toward greater transparency and accountability in the management of their oil revenues. The whole sorry saga had started in the mid-1980s, when a sudden drop in the price of crude oil precipitated a cycle of debt (all too familiar to African oil producers) from which the country has never extracted itself. Unable to pay off its external creditors in the late 1980s, the Angolan government found its once-excellent credit rating in the gutter and was forced to ask the IMF for help. The IMF doesn't reschedule a country's external debt without first requiring the country to meet stringent conditions (usually to do with reining in government spending). The Angolans, accustomed to no-questions-asked cash from the Soviet Union, balked at the requirements laid out by the IMF and negotiations stalled. But as the 1980s drew to a close, the Soviet Union began to implode, and the MPLA government had to find ever-more-creative ways to raise cash for its expensive war against the UNITA rebels (who still received heavy financial backing from the committed Cold Warriors of the Reagan administration).

At this point in the early 1990s, the ideological bankruptcy of the Cold War proxy battle over Angola reached its farcical extremes. Looking to replace financial and technical support from the now-collapsed Soviet Union, the ruling MPLA, founded by Marxist revolutionaries as a liberation movement fighting Portuguese colonialism, found itself increasingly reliant on revenues from its offshore-oil exploration. Ironically, this work was conducted—as it still largely is—

by the American oil company Chevron and the French state-owned firm Elf (since privatized and renamed Total). The MPLA also increasingly applied to private Western banks for credit, using sales of future crude-oil production as collateral. At one point, the Angolans were receiving hundreds of millions of dollars from the United States Export-Import Bank, a quasi-official export-credit agency and loosely regulated arm of the U.S. government, which by this point had switched its allegiance from UNITA to the MPLA. Thus, abandoned by their Soviet paymasters, a band of lapsed Marxists turned to Western capital markets to pay for a war against pro-Western rebels, who were in turn dropped by the West. With the fig leaves of liberalist and communist ideology gone, Angola descended into the bloodiest phase of the conflict, with an estimated 300,000 people dying in the war's final years.

However darkly comic it may appear in hindsight, the increasing reliance of the Angolan government on commercial oil-backed loans marked a serious departure from the normal channels of international financing for an African government, and one that would have devastating consequences in later years. Unlike traditional loans by rich nations to poorer ones, oil-backed loans raised through the capital markets have strict repayment terms and steep interest rates, and leave the debtor highly vulnerable to fluctuations in the price of oil. Indeed, more than any other single factor, it has probably been the price of oil that has determined the tone of relations between Angola and the IMF. In the early 1990s, as the first U.S.–led war on Iraq pushed up oil prices, the Angolan government benefited and felt less of a need to waste its time negotiating with the IMF. But by the mid-1990s the price of oil was in free fall, forcing the government into another cash crunch. In 1995 a Staff Monitored Program (SMP)—a precursor to a fully-fledged IMF debt-readjustment package—was agreed to by the MPLA, but abandoned after a few months, largely because the government was reluctant to allow the IMF access to its records. In 1998 another SMP was negotiated, but not signed by President dos Santos.

By the end of the 1990s, oil prices had hit record lows, and UNITA had launched a fresh offensive in the country's interior. The MPLA had its back against the wall. A third SMP was agreed on, and this time ran from 2000 to 2001 before it, too, collapsed in the face of rising oil prices. This time the issue was the IMF's insistence that the Angolan government account for the billions of dollars that appeared to be missing from the national treasury.

During a visit to Washington in February 2002, dos Santos voiced his irritation with the IMF's dogged pursuit of revenue transparency, telling the Voice of America that "the police action of the IMF" was unacceptable and that the Fund "should respect the sovereign rights of the Angolan state." The timing of dos Santos's most forceful public condemnation of the IMF was no coincidence. The September 11 attacks were just months old, the price of oil was climbing quickly, and with African oil the subject of growing interest to American lawmakers, dos Santos had been invited to the United States. His comment signaled that if the United States wanted Angola's oil, it would have to take steps to rein in the IMF. Dos Santos held all the cards, and he was happy to let his American hosts know it.

Since the last Staff Monitored Program broke down in 2001, the Angolan government has discovered it has less and less use for the IMF and even less patience for its lectures about transparency. International oil prices have set record after record, Angola's oil production has nearly doubled, and in 2004 the Chinese put a sweet frosting on the cake in the form of a $2 billion credit facility (later increased to $4 billion) in exchange for a lucrative oil-exploration license. No amount of Western donor aid or IMF debt rescheduling can compete with the kind of money that has been flowing into the Angolan treasury since 2002, whether from China or from ExxonMobil. In 2005 Angola received twenty times more oil revenue than foreign aid. At press time, any agreement on a new SMP seems very much on ice, and the Angolans have continued to use oil-backed loans from commercial banks to help bankroll their massive postwar reconstruction effort.

However, it would be a mistake to conclude that the Angolan government no longer "needs" the IMF or Western donor money. An IMF imprimatur on the country's finances would amount to a critical seal of approval for the MPLA and one that would attract far greater foreign investment into the country. Western donor nations have remained largely unamused by reports of billions of dollars in disappeared funds, and have all but ruled out the possibility of a postwar donors' conference for Angola. This has been a source of great bitterness in Angolan government circles, where leaders have watched as first Afghanistan and then Sudan were lavished with billions of dollars pledged by donors at similar conferences.

The series of damning reports by Global Witness and the torrent of negative stories that followed in the international press caused much anger in Luanda. But, as embarrassing as Global Witness's accusations may have been for the Angolan government, they are potentially more troubling for the Western oil companies, which must face activists and shareholders in their own countries. Global Witness is just one of nearly three hundred NGOs around the world that have come together under the banner of a movement called "Publish What You Pay," which aims to make international oil companies legally obliged to make public their payments to the governments of countries in which they operate. PWYP campaigners argue that such transparency would make it harder for governments to hide corruption.

Perhaps not surprisingly, oil companies have been slow to warm to Publish What You Pay, arguing that publishing such sensitive data would put them at a competitive disadvantage against rivals from countries without PWYP laws. After all, if you are a corrupt government, who would you rather do business with: a company that makes its dealings with you a matter of public record, or a more discreet partner? "I consider PWYP a little bit unilateral," Fernando Paiva, Chevron's head of policy, government, and public affairs in Luanda, told me when I caught up with him at the company's headquarters (which sits on the corner of Lenin Avenue and Salvador Allende

Boulevard). "You would think that it should be the responsibility of the government. They should publish what they *receive,* and this is what they do when they publish the annual budget." Like many oil executives, Paiva believes the Extractive Industries Transparency Initiative (EITI), a pet project of British Prime Minister Tony Blair, is a "more realistic" approach. Launched by Blair at the World Summit on Sustainable Development, held in Johannesburg in 2002, EITI is a set of operating principles designed to promote transparency in the use of oil and mining revenues—one that both governments *and* oil companies can sign on to as a way to demonstrate their bona fides.

In recent years, EITI has gained far more momentum and traction than PWYP, in no small measure because the oil companies have deftly managed to water down most of EITI's original stipulations. In 2003 the *New York Times* reported how months of behind-the-scenes negotiations between American oil companies and Bush administration officials resulted in the British government being forced to abandon some of its more ambitious proposals, such as the signing of a compact requiring detailed reporting by oil companies. In the end, EITI placed considerably more of the transparency burden on host governments than the British government had originally envisaged. According to the *Times,* the most significant impetus behind EITI's evisceration was ExxonMobil.

Nevertheless, whatever its imperfections, few disagree that EITI is a step in the right direction. From the perspective of the oil companies, signing on to EITI principles makes it possible to deflect accusations of complicity with government corruption, and absolves them from making gestures that might endanger their operating agreement with the host country. In Angola especially, the fear of acting unilaterally, as Publish What You Pay recommends, when billions of dollars are at stake is not unjustified. In 2001 the British supermajor BP, thinking it was offering the world a demonstration of its good faith, took the unusual step of making public the signature bonus it had paid the Angolan government for a new oil concession, which drew a ferocious response from Angolan authorities. Manuel Vicente, CEO of the state oil company Sonangol, wrote a stinging letter of rebuke to BP, threatening to cancel all its contracts in Angola if it continued to publish such data. The letter accused BP of "seriously violating the conditions of legal contracts signed with Sonangol" in an effort to "attract bogus credibility" from certain "organized groups [engaged] in an orchestrated campaign against Angolan institutions," and, in closing, stressed that "we strongly discourage all our partners from similar attitudes in the future." Vicente saw to it that copies of the letter were sent to every oil company operating in Angola.

Global Witness, the IMF, Publish What You Pay, and the international transparency "agenda," as the Angolan government sees it, are delicate subjects in Luanda, and not the easiest of topics to bring up in conversation with government officials who are keen to avoid inflicting further damage on the country's image abroad. A variety of defenses and rebuttals are offered in response to the suggestion that Angola has an endemic problem with corruption and mismanagement of oil revenue, chief among these being the issue of institutional capacity. After a decades-long war that left the country's infrastructure in tatters and the national bureaucracy barely able to function, it should not be surprising, it is argued, if errors of accounting appear, or if the state lacks the legal, fiscal, or bureaucratic mechanisms to prevent unscrupulous people from dipping into the honeypot from time to time. After all, when the Portuguese left in 1975, they left abruptly, and took 340,000 settlers with them—instantly robbing the country of its technocrats and managers. Since civil war broke out immediately after the Portuguese departure, the newly independent Angolan state never had the chance to develop its own cadre of trained technocrats. By 1998 only 3 percent of government employees even had a university education.

Although it may seem unfair to hold Angola to the same standards of accountability and governance as a peaceful, developed, and fully democratic Western country, transparency advocates insist we must be careful about allowing sympathy for the country's brutal history to color our judgment. Accounting is not rocket science. Or, as Global Witness put it, "a government and state oil company that handle billions of dollars through complex arrangements, including the use of Special Purpose Vehicles and foreign tax havens, can certainly manage a simple balance sheet." Indeed, Sonangol is run as efficiently and as professionally as any multinational oil company.

Far more convincing than the institutional-capacity argument, and ultimately far more honest, is the fact that a vast and highly underqualified postcommunist bureaucracy of MPLA loyalists accrued during the government's long battle against UNITA. Some of these "dinosaurs" have been in their jobs for decades, and many even played an active role in the liberation struggle of the 1960s and 1970s. As the civil war came to an end, many of them felt the state owed them something for their loyalty to the cause. "Corruption in Angola is not on the scale suggested by Global Witness and others," said José Oliveira, editor of *Revista Energia* (*Energy Review*), from his crumbling stucco office on Luanda's waterfront. "It is not institutionalized as it is in Nigeria, say. It is more the case that you have these guys who have fought for years for the MPLA and have been abandoned even without the prospect of being able to collect a retirement later on. When the means to deliver has not been there, they have started helping themselves." They are allowed to continue on, because the younger and more reform-minded leaders in the MPLA cannot risk the bad blood that would come from a mass sacking of regime loyalists. Many of these people, after all, have been fighting imperialism since before their fresh-faced young paymasters were even born.

◂◂▸▸

"Global Witness is a private enterprise. They can be paid to write anything against us," said Deputy Foreign Minister Jorge Chicote when I met him in the drawing room of the peach-colored Portuguese-era Ministry of Foreign Relations building in Luanda. Chicote is a short, round man, not much older than forty, with an air of extreme wealth about him—the kind of man who probably sleeps with a Portuguese passport under his pillow. He speaks in a milk-smooth stream of magisterial English with a slight American twang and can be disarmingly honest about what he thinks. When he had finished telling me how Global Witness had been brought in by opposition politicians in Angola as part of an effort to tarnish the MPLA, he proceeded to suggest—elliptically but unmistakably—why so much money was unaccounted for from the years 1997 to 2001. "There was an arms embargo," he reminded me. "We had to find a number of unconventional ways of acquiring equipment for our defense."

Of all the explanations you hear for the "missing billions" in Luanda, this one is not only the most convincing, but also the one the Angolan government is least willing to be second-guessed on. During the final years of the civil war, UNITA was clearly losing on all fronts, but had grown more entrenched and more desperate, resorting to guerrilla tactics, scorched-earth policies, and terrorizing civilian populations. The MPLA maintained they needed to bring the war to a decisive end, and the only way to do that was to overwhelm UNITA with a vastly superior display of firepower. To do that, they needed large sums of money and they needed it fast, and it didn't matter where it came from. If they had played by the rules of the international community, they insist, the war would have simply dragged on and thousands more people would have died.

As it was, the international community had already let them down in spectacular fashion.

In the 1970s and 1980s, when Angola was an important pawn in the Cold War battle between communism and capitalism, the Soviets were only too happy to shower money and weapons on the MPLA, while the Americans and South Africans did much the same for UNITA (who at one point were receiving so many planeloads of military equipment that they couldn't distribute it fast enough to their troops). But when both white-ruled South Africa and the Soviet Union became consumed with their own dramatic internal changes, the United States sensed an opportunity to bring the war to an end, and UNITA to power. In 1991, supported by a largely symbolic "troika" of the willing composed of the sycophantic former colonial power Portugal and a much-emaciated Russia, the United States brokered the Bicesse Accords, which, with hindsight, have gone down in history as a catastrophic disaster for Angola. An unrealistic early election date was set, with the MPLA allowed to maintain control of the transitional government until then (so that UNITA could not be held responsible for any failure to deliver). The UN's role and funding was kept to a minimum to avoid gumming up the works or otherwise complicating the assumption of power by UNITA. Dame Margaret Antsee, who was in charge of the UN mission at the time, has described her attempts to fulfill her mandate with an $18.8 million budget as like "flying a 747 with only enough fuel for a DC-3."

The Americans' rough-and-ready strategy to use the end of the Cold War as an opportunity to catapult their allies into power might have gone off smoothly had it not been for one minor hitch: The majority of the Angolan people failed to vote for UNITA. In the legislative elections of 1992, the MPLA won 54 percent of assembly seats while UNITA secured only 34 percent. In the presidential contest, dos Santos won 49.6 percent of the vote to UNITA leader Jonas Savimbi's 40.1 percent. The UN declared the elections "generally free and fair" but UNITA, led to believe they were about to take power in Luanda, rejected the results and returned to fighting. The two years of warfare that followed were some of the worst the country had ever seen. Some 300,000 people lost their lives as whole cities were reduced to moonlike layers of rubble. Following the terms of Bicesse, the MPLA had begun to demobilize government forces, and this gave UNITA a strong tactical advantage. The UNITA position was further strengthened by renewed support from its old ally Mobutu Sese Seko of Zaire, by the global proliferation of illicit light weapons following the breakup of the Soviet Union, and by the consolidation of UNITA's control over Angola's lucrative diamond mines. The MPLA, however, still controlled the country's offshore-oil production; the Gulf War drove up oil prices and made it possible for the MPLA to buy more guns and cancel out UNITA's military advantage. After years of being told their country was the scene of an epic clash of social systems and beliefs, the Angolans now saw their conflict descend into a brutal arena of personal greed and ambition, reduced to the headline OIL VS. DIAMONDS.

UNITA leader Jonas Savimbi had become an embarrassment for the Americans. Driven by what one historian has called a "messianic sense of destiny" and others have described as "psychopathic tendencies," he began building his own increasingly bizarre cult of personality in the middle of the jungle and made it clear that UNITA was digging in its heels. Faced with such barefaced recalcitrance and no longer feeling threatened by the Soviet menace, the United States abruptly switched its allegiance to the MPLA, which in 1990 had shrewdly dropped its ideological commitment to Marxism-Leninism. So complete was that transformation that in 1995, when the UN General Assembly voted to condemn the American blockade of Cuba, the Angolans abstained. A country that had once played host to 50,000 Cuban troops fighting the South African Defence Forces, and that still sported a national flag based on the hammer and sickle, was now officially in bed with the *Yanquis*.

In 1994 the world made another vain attempt to impose peace on Angola, this time in the form of the Lusaka Protocols, signed in the Zambian capital and presided over by Zambia's President Frederick Chiluba. The leaders of UNITA and the MPLA showed up for the photo op, but it was clear even at the signing ceremony that there was no genuine appetite for peace on either side. Savimbi, who had spent much of the past twenty years in the bush, turned up dressed like a backup singer for the Temptations, in a broad-collared suit and no necktie, looking like the cat that ate the canary and still got a seat at the table. Dos Santos, crisp and presidential in a trim tailored suit, appeared stern and unamused, with an expression that seemed to ask, "Does anyone seriously expect us to trust them this time?"

Indeed, UNITA leaders dragged their feet implementing the terms of the protocols, and the international community became deeply frustrated. For the first time, the world unanimously chose to treat the MPLA as the only legitimate Angolan government, forcing UNITA to fall back on its diamond revenue to keep the conflict alive. In the tense period that followed from 1994 to 1998, Angola was neither at war nor entirely at peace. Zairean dictator Mobutu Sese Seko, who for years had been a loyal friend of both the United States and UNITA, was overthrown, and an international ban on uncertified Angolan "conflict diamonds" came into effect, cutting off UNITA's last source of revenue. Dos Santos saw his chance and seized it. At an MPLA party congress in December 1998, he said the only path to true and lasting peace led through war, declared Lusaka dead, and began a major offensive against UNITA.

Over the next two years, UNITA reverted to guerrilla tactics, ravaging and displacing rural populations in an effort to keep from starving to death in the bush. For its part, the MPLA intervened in the civil wars raging in both Zaire and the Republic of Congo, hoping to install friendly regimes and cut off UNITA's supply lines and close down their rear bases—a bit of interference to which the UNITA-weary international community was happy to turn a blind eye. In early 2002 the Angolan Armed Forces closed in on UNITA's final command base in Moxico, killing both Savimbi and his vice president, Antonio Dembo. What was left of the UNITA leadership came in from the bush, malnourished and emaciated, with no choice but to sue for peace on terms so highly favorable to the government that they made the resumption of war an impossibility. After years of playing by the limp-wristed rules of the international community, the Angolans had brought the conflict to an end in the only way that made sense after decades of bloodshed—with an overwhelming display of military force.

This, at least, is the MPLA's version of events, and it is one with which several senior Western diplomats I spoke to in Luanda were cautiously sympathetic. "To run a war in Africa, you need money quickly," one of them told me. "Leading up to 2002, dos Santos wanted the war to stop, using the military option. There was an arms embargo, and they feel it was absolutely justified. They needed money in bags and suitcases, and they feel that if there had been transparency, they wouldn't have been able to bring the war to a close."

Another was even more succinct: "They don't feel they owe much to the West," he told me. "They feel that the international community let Angola down; that they could have brought the war to an end sooner and didn't; that they weren't there for Angola. To a certain extent, that underlines their attitude to Western financial institutions today. There is an unwillingness to be forced into doing anything by the West."

Why give the international community so much control over your economy when they have done so little to earn your trust? Why accept that technocrats in Washington know more than you do about fiscal management when technocrats in Washington have thus far proven capable only of helping the country's civil war drag on for ten years longer than it should have done? It is bad enough what you have done to our garden, says the MPLA; now you want the keys to our house?

This is part of a larger anti-imperialist rhetoric that has been adopted by the Angolan leadership over the years—both in private conversations and in public speeches—and it plays well in many quarters. When the Americans recently tried to nudge the MPLA into holding elections (there have been none since the ones in 1992 that reignited the fighting), President dos Santos had a clear message for them and to the world at large: "Democracy," he said in his birthday address to the nation in 2005, "was imposed on Africa by the West," and his comments drew quiet nods from many on the continent.

At the root of this refusal to be dictated to is the fact that—between the vast amounts of cash and military support received from the world's superpowers and its oil and diamond patrimony—Angola is one of the few African countries that have never been reliant on foreign aid for economic survival. Faced with threats about the annulment of loan guarantees or the cancellation of development aid, virtually every African head of state has eventually had to make at least superficial attempts to be seen playing by the West's rules. But to the MPLA, the concept of IMF "conditionality" is not only unfamiliar, but an affront. One Western diplomat and veteran Africa hand told me, "People who are used to dealing with aid-dependent countries get a big shock when they first come here."

The divine intervention of a booming offshore-oil industry in the midst of a long period of high oil prices has only added to Angola's self-regard as the one country that stood up to the white man and got away with it. And a spectacular win on the soccer pitch hasn't hurt. But as much as these events have fed into the country's narcissism of late, they are not sufficient to explain why the Angolan government has dragged its feet so much on making and meeting transparency and clean governance commitments to the IMF.

There has also been the small matter of a certain injection of Chinese cash.

◄◄·►►

CHINESE CREDIT LOOSENS UP THE BANKS, crowed the front page of the state-controlled *Jornal de Angola* on the day I landed in Luanda. It was October 1, 2005, nearly a year after the fact, but Angolans were still slapping themselves on the back. In late 2004 China's state-owned oil company, CNPC, facing stiff competition from its Indian counterpart for the rights to a lucrative deepwater-exploration license, had decided to nip negotiations in the bud with a $2 billion loan guarantee from the Chinese government. Needless to say, China's heavy-handed strategy had worked, and the Indians had woken up to find the game over before it had even begun. For the Chinese, this latest move in an aggressive push into Africa was part of a major national objective to secure a supply of oil for the country's blossoming industrial economy. But from the Angolan perspective, it meant $2 billion to spend on postwar reconstruction without the humiliation of going cap in hand to the IMF. It was a major political victory for the MPLA and, even a year after the fact, they were still gloating about it. Under the banner headline was a picture of José Pedro de Morais, Angola's relatively reform-minded finance minister, looking smug, and inside were several pages of an interview in which Morais explained how the Chinese loan had made it possible for Angola to obtain better borrowing terms on the international market, and all but declared the IMF an irrelevance. "Our main objective," he told the interviewer, "has been to redefine the work equation with the IMF."

But even the Chinese don't lend a country $2 billion without laying out some terms and conditions. Unlike the IMF, though, which drones on earnestly about transparent bookkeeping and sound fiscal management and priority sectors, the Chinese had a slightly cruder condition for their credit line: 70 percent of the contracts for reconstruction projects that the loan financed would have to go to Chinese companies. Everyone from the IMF to Global Witness to Western embassies furrowed their brows at such a crass and old-school approach to development aid and expressed concern about what the lack of "Angolan content" would mean in a country where local people desperately needed jobs. How would a traumatized population react to the sight of thousands of Chinese laborers building railways?

But the Angolan government dismissed Western expressions of concern as hypocritical and thinly veiled manifestations of sour grapes. Where were the knitted brows, they asked, in the 1970s, 1980s, and 1990s, when Chevron made a fortune offshore of Cabinda while a civil war raged? Do Western countries not put pressure on recipients of donor aid to ensure that the ensuing contracts go to their companies? Was Halliburton not rebuilding Iraq? No, the Angolans were going to build their roads and railways, and it didn't much matter where the money came from or who did the labor.

One American diplomat I spoke to expressed grudging sympathy for the pragmatism of the Angolan government. "Look, let's be honest, I didn't give them the road, did I?" she told me. "I didn't give them a new airport, or railways." The sentiment was one I heard again and again in Luanda's Western chanceries. On the record, Western diplomats were quick to voice concern about Angola's embrace of the assertive Chinese presence, but off the record, most confessed, who could blame the poor bastards? They were getting a shiny new airport, a railway link to Zambia, a 125-mile road up the coast, and a bunch of new government buildings, and all without having to open their books and explain to the world what happened to the $4.2 billion that disappeared in the late 1990s. With a little help from the Chinese, with a little help from global events that have pushed up the price of crude, and with a few risky oil-backed loans, the Angolans have, for the moment, won their battle with Western governments.

Some of Angola's most-traditional allies have felt most acutely the distance the government has tried to put between itself and the West. France, whose influence in the subregion was once unrivaled, has been cold-shouldered by the Angolans following a French court's decision to prosecute Pierre Falcone, a notorious arms trafficker with close ties to President dos Santos. In 2004 the new French ambassador, Guy Azais, was made to wait more than six months after his arrival in Luanda before being invited to present his credentials to the president. Meanwhile, the French oil giant Total found itself unable to renew a long-standing drilling license, which Sonangol eventually awarded to a Chinese company. And the ultimate indignity came in early 2005, when Air France had its request for a second landing slot in the lucrative Paris-Luanda route denied, with the privilege given to British Airways instead. "We have a difficult relationship with France at the moment," admitted Jorge Chicote, Angola's deputy foreign minister, when I asked him about the Falcone affair and its fallout.

But relations with the former colonial power Portugal have been only slightly more cordial. During my visit to Angola, the papers were full of nasty accusations by the authorities that Angolan citizens were being "humiliated" when applying for visas at the Portuguese consulate in Luanda. And Portuguese businesses, which for years treated their bloated contracts with the Angolan state as a birthright, have suddenly found themselves unable to compete with Chinese firms that will do the same work twice as fast for a fraction of the cost. In 2005 and 2006 a new 230-mile road was being built from Luanda to Uige province by a Chinese contractor charging a mere $211 million. By way of comparison, a Portuguese firm was simply adding two lanes to a short 16-mile stretch of road between Benguela and Lobito for a hefty $24.2 million. But even more striking than the price comparison is the fact that both projects were going to take about the same length of time to complete. "There is simply no competing with the Chinese," sighed one Portuguese banker I spoke to, with a shake of his head.

As if to illustrate the point, the Chinese were everywhere in Luanda. Supermarket shelves, usually laden with Lebanese and either Portuguese or French products in this part of Africa, now also displayed cheap Chinese crackers and cans of vegetables from China. The hotel lobbies were filled with potbellied Chinese businessmen, milling about, looking open for business and ready to inherit the earth. And when I visited the gleaming new Ministry of Finance building in downtown Luanda, haggard-looking Chinese contractors, cheap unfiltereds dangling from downturned lips, were putting the finishing touches on the glass-paneled salmon-colored edifice. The building itself looked like a cheap Chinese knockoff of a Western office block. Clumps of exposed wire were still dangling from panels where the intercom was going to go. In the elevators, the buttons were labeled only with Chinese characters.

Despite the assertive and highly visible new presence of the Chinese in Angola, it would probably be an oversimplification to say, as many international press reports have done in recent years, that the Chinese loan has been a "victory" for the MPLA or that it has allowed it to "dodge" the scrutiny of the IMF and the international community. This has been a line of analysis particularly beloved of the American press, which has tried to fit it in with a larger narrative of yellow peril in which Beijing, unconcerned with human rights and good governance, "cozies up" to "rogue regimes" in Africa in an effort to secure oil supplies, thus scuppering efforts by the West to bring these regimes "into line" with international norms. Though there may be an element of truth to this, in the Angolan case, at least, the MPLA would still dearly love to reach an agreement with the IMF as, without one, it will face real difficulty attracting crucial foreign investment into the country. "I think they're aware that no amount of Chinese credit will have the same effect as having an agreement with the world economic system," the U.S. diplomat told me.

Even if one accepts that there is a trend in Luanda toward replacing traditional Western alliances with Asian ones—a "look East" policy not unlike that of Zimbabwe—one would also have to acknowledge that there is at least one glaring exception to the rule, and it is the United States. The Americans, once bitter enemies (not to mention active underminers) of the MPLA, have enjoyed a genuine renaissance in their relations with Angola since President dos Santos was welcomed to the White House in 2004 as an important ally in the Global War on Terror. (Angola had been on the UN Security Council in 2003 and, still feeling insulted by the French pursuit of Falcone, had been happy to vote with the Americans.) "Even if the past has been difficult, relations with the U.S. have been extremely good in the past two, three, four years," says Chicote. "America is a priority partner for our international cooperation."

During my three short weeks in Angola, I saw no shortage of evidence for this budding U.S.-Angola friendship. Operation Med-Flag, a series of joint exercises between U.S. and Angolan armed forces, had just wrapped up, and Senator James Inhofe, the influential chairman of the Senate Armed Services Committee, made time during his African tour to stop in Luanda—ostensibly to thank the government for its participation in Med-Flag, but also to shoehorn in a few hours of talks with Angolan officials about stepping up U.S. military coop-eration. But the most tangible symbol of the special relationship was the gargantuan U.S. Embassy building that had just been erected on a hilltop in the posh Miramar district, at a cost of $70 million. The 54,000-square-foot building, its six-foot-thick walls decorated with red-white-and-blue bunting, was visible from almost anywhere in downtown Luanda, and, perhaps most important, dwarfed the far-more-understated French Embassy nearby.

As a visiting U.S. journalist, I was invited to the opening cere-mony for the building on the morning of October 13, 2005. After nav-igating a maze of concrete barricades, bulletproof windows, and metal detectors, I was ushered into the garden by Phil Nelo, an Angolan press handler who spoke English with a bright American twang that he peppered with expressions like "thanks a bunch." On a table were three enormous cakes—one iced with an American flag, one with an Angolan flag, and the third shaped like the embassy building itself. As she cut the ribbon, the U.S. ambassador, Cynthia Efird, praised the "winning team" that had made the building possible, and called it "a symbol of the strong relationship between the U.S. and Angola." Efird referred to Angola as a "friend" of the United States, stressing that her use of the word was not accidental. "It symbolizes my government's relationship with Angola," she told those assembled.

Cynics might say it symbolizes her oil-addicted government's desperate need to win friends in countries with formidable petroleum reserves. A country that will soon be producing nearly as much crude as Nigeria, America's fifth-largest supplier, and (unlike in Nigeria) all of it safely offshore, is hardly one the United States can afford to ig-nore. But American officials are adamant that their newfound love for the MPLA has little to do with petro-politics, pointing instead to An-gola's growing profile as an African economic powerhouse and U.S. trading partner, as well as the fact that the size and skill of its army allows it to play an important "stabilizing" role in the neighboring

Congos. "Angolan oil will have peaked by 2012," an American diplomat reminded me testily. "But, as this building shows, we're here to
stay for the long term." All this may be true, but it will have escaped
no one's attention in Miramar that in 2005 China overtook the United
States as the largest importer of Angolan crude.

But even with all the friends in the world—American, Chinese, or
otherwise—Angola faces formidable challenges in the years ahead,
beginning with the nation's devastated physical infrastructure. Outside Luanda, roads and bridges are virtually nonexistent, and even in
the capital, the traffic lights left behind by the Portuguese haven't
worked in years. (Cars actually speed up, rather than slow down, as
they approach signalless four-way intersections, in an effort to make
it through before the cross traffic cuts them short—a particularly har-
rowing game of chicken even for the most steel-nerved of passengers.) And even as the government prepares to register citizens for the
long-awaited elections to be held by the end of 2007, it has had to
contend with the fact that hundreds of thousands of refugees have
begun returning home since the war ended, most of them without
documents and many of them to a country they barely remember, or
have never seen.

◄◄--►►

Of all these challenges, though, the one rarely mentioned—by the
Angolans or by the international community or by the oil companies
or even by the international press—is the long-simmering guerrilla
war in the breakaway province of Cabinda. The conflict has proven
even more intractable than the Angolan civil war, and goes straight to
the heart of the country's rising fortunes as a global oil player.

Cabinda, a tiny sliver of land the size of Puerto Rico, is separated
from the rest of Angola by an even tinier sliver of the Democratic Republic of Congo (formerly Zaire). Unlike Angola, which was a Portuguese colony since the fifteenth century, Cabinda was a triumvirate

of kingdoms (Kakongo, Ngoyo, and Loango) until 1885, when the three kingdoms sought Portugal's protection against the Belgian and French Congos on either side of them and became a Portuguese protectorate sometimes known as the "Portuguese Congo." When the Portuguese abandoned their African possessions in 1975, Angolan soldiers occupied Cabinda immediately and declared it part of Angola. No one knew that billions of barrels of oil lay off its coastline, but the land grab proved to be one of the smartest things the MPLA ever did. Today Cabinda is home to barely 2 percent of Angola's population and represents less than 1 percent of its total land area, but accounts for 60 percent of the country's oil production.

Culturally and linguistically, most Cabindans are Kongo people of one kind or another—Portuguese-speaking only by an accident of history and closer to their Francophone neighbors in the two Congos between which the province is wedged. Throughout the Angolan civil war, the MPLA was keenly aware of Cabinda's potential as a rear base for attacks or supplies from either the Republic of Congo or from Zaire's Mobutu, whom they always suspected of wanting to annex Cabinda for himself. As the province's oil potential grew, it became an important source of revenue for the MPLA, and they fought even harder to retain control. Some of the civil war's most brutal battles were waged between the MPLA and FLEC (Cabinda Enclave Liberation Front), the rebel group fighting for Cabinda's independence. FLEC was never included in any of the peace negotiations, from Bicesse to Luena, and it has subsequently never given up its battle against the MPLA. When I visited Cabinda, an unknown number of Angolan troops—believed to be between 30,000 and 50,000— were still stationed, and fighting continued in the mountainous Maiombe forest. Given the enormous stakes involved, the Cabinda conflict was an issue of extreme sensitivity to the Angolans and one they were going to great lengths to keep out of the public spotlight. The government considers the conflict an "internal matter" and has long made it clear that it welcomes neither international mediation nor the prying eyes of foreign journalists. As a result, many Africa analysts consider Cabinda one of the world's most underreported conflicts.

Before I traveled to Cabinda, I was warned by everyone I spoke to that I should at all costs avoid staying at the Hotel Maiombe—the only decent place in town—because it was a "den of spies," where I could expect not only phone taps but also hidden cameras and microphones. It seemed advice worth heeding. On arrival, I was met at Cabinda's airport (a landing strip and an open-air receiving pen with a corrugated tin roof) by João Conde, a local activist who had been born in 1978 in a Congolese refugee camp and now taught philosophy and Latin at the local high school. Conde took me a couple of miles outside town to a hotel he assured me would be much mellower than the Maiombe. Even there, though, as soon as we had sat down to chat on the terrace, his eyes began to dart back and forth and he grew nervous.

I later learned that a large man in a baseball cap who was sitting behind us had been listening closely to our conversation. Throughout my time in Cabinda, the large man in the baseball cap was never far away, popping up with friends at a nearby table every time I sat down for a drink with someone, or keeping a few steps behind me as I wandered the streets in town. Barely 200,000 people live in all of Cabinda, many of them in remote mountain villages, so in the small cluster of streets and businesses that make up Cabinda city, few activities go unnoticed for long.

In this respect, Cabinda is no different from any sleepy small town anywhere in the world. But Cabinda is a sleepy small town with a difference, and the difference has to do with the multibillion-dollar oil industry that operates in its territorial waters. In 2000 tax receipts from Cabinda to the Angolan National Treasury totaled $2.5 billion— 71 percent of government oil revenue. Luanda allows the province to retain 10 percent of its oil revenues, but Cabindans look around at their poor living conditions and wonder where this money is going. Unlike in Nigeria, where state governors are elected and at least generally come from one of the local tribes, Angola's provincial governors and technocrats are appointed, in classic top-down Marxist fashion, in Luanda. In Cabinda they are seen as interested in little more than maintaining order and undermining separatist activity.

If Cabinda were allowed to become an independent nation, its citizens would be among the richest in the world, and it is hard to imagine that this has crossed no one's mind here. Nevertheless, Cabindans swear up and down that their struggle is entirely a cultural and historical one. "They can explore all the oil they want in the ocean," says João Conde, reflecting a common point of view here. "Just let them leave us in peace to run our own affairs." And it is true that Cabinda's struggle against Angolan rule began well before large-scale oil exploration got under way off its coast. FLEC was founded in 1963 as a guerrilla movement aimed at securing Cabinda's independence from Portugal and feels it has been fighting the same anticolonial battle ever since—only the enemy has changed. Cabindans are fond of invoking the spirit of the Simulamboco Treaty of 1885, when their three ancient kingdoms agreed to allow Portugal to act as their protector. Cabinda's status as a protectorate distinguished it from the full-scale colonial project the Portuguese ran in Angola, and many feel it was the Angolans who first "colonized" them in 1975, sending thousands of soldiers and functionaries to live among them and eventually changing the ethnic makeup of Cabinda. To this day, the anniversary of Simulamboco is celebrated spitefully in Cabinda, and an old stone monument to it on a hill outside town has become a revered symbol of Binda nationalism.

If push came to shove, most Cabindans would probably accept a compromise arrangement under which the province received autonomy and a larger share of oil revenues, rather than full independence. But the Cabindans claim that Luanda is not willing to discuss the subject. "As long as they have their war machine here," says Manuel Gomes, director of a respected local NGO, Gremio ABC, referring to the Angolan troops in the province, "they will never find a solution."

Gomes, who plays by the rules and is grudgingly tolerated by the authorities, told me the Angolan government's mistake has been to treat the Cabindan problem as an offshoot of their civil war, thinking that a crushing military defeat of FLEC would bring an end to the conflict. But FLEC is not UNITA, he cautioned. UNITA owed much of its survival to external support from the United States and others and, especially in the later years, aroused very little public sympathy. "Beating UNITA was just a question of getting rid of Savimbi," he said. "The problem of Cabinda is very different: You have to deal with the people of Cabinda."

Gomes's work dealt mainly with the environmental impact of oil exploration on host populations, and even a quick look around showed he had his work cut out for him. One of the first things I did upon arriving in Cabinda was to walk down to the beach to look out at the line of flaring oil platforms and hulking tankers parked a few miles out to sea. The beach was a tropical cliché, framed in palm fronds and speckled with tiny shells and naked gamboling children, but there was a dissonant note amidst the breezy, sun-drenched magic. The peach-colored sands were marred by long, inky streaks of jet black. When I rubbed the sand between my fingers, it felt unmistakably oily.

A Chevron spokesman told me that the "dark-colored sediments" on the beach at Cabinda were mostly the result of "naturally occurring sources" and did not present an environmental or health risk. He cited a 1997 study commissioned by the company that indicated that the amount of petroleum in the sediment was "low" and "not of a magnitude to be considered an environmental concern." In other words, it's just black particles washed up naturally by the Congo River.

But Gomes didn't buy it. "The Congo has been here for centuries," he told me. "Ask anyone who grew up here, and they'll tell you no, the beach was never like this."

Chevron's credibility in Cabinda now seems lower than it has ever been, and that has to do with more than just its explanation for why the beaches are turning black. At the core of local frustration with the company is Malongo, its operating facility on the outskirts of Cabinda city. Originally built by Gulf Oil (most of which Chevron acquired in 1984), Malongo was surrounded at the height of the civil war by a double layer of electrified fencing and razor wire as well as several hundred landmines. In Angola's absurdist version of the Cold War, the mines were laid by Cuban soldiers under orders from the MPLA to protect Gulf Oil's operations against UNITA. In other words, for years in Cabinda, a revolutionary Marxist government depended on money from an American oil corporation whose operations were defended by Cuban soldiers against attacks by an American-backed rebel army.

Today, despite the end of active warfare (in July 2006 the Angolan government signed a peace accord with FLEC, though several militant factions rejected the accords as illegitimate), the fortifications remain and Malongo is seen by locals as an eyesore if not a menacing forbidden city of Western capital that contributes little to the local economy. In the past, FLEC guerrillas have kidnapped foreign oil workers passing through town, so Chevron's American and expatriate personnel are picked up in buses when they arrive at Cabinda airport and taken straight to Malongo, from whence they are flown by helicopter to the offshore rigs for their thirty-five-day shifts. At Malongo they enter a cocooned existence of satellite TV, basketball courts, imported American snacks, and even a rolling golf course, all designed to discourage them from leaving the compound.

When I visited in October 2005, the company's Cabindan employees had just returned to work following a wildcat strike that had lasted several days. The Cabindans claimed they were discriminated against by Chevron, which preferred to hire "expatriates" (i.e., Angolans) whom they knew would not be members of the banned activist group Mpalabanda. Several Malongo employees I spoke to said that if they had Congolese qualifications, for example, they were automatically suspect, because most radical Cabindans had spent many years in Congo. Ultimately, the strike had ended when the Angolan army stormed Chevron's compound. It was a particularly clumsy attempt by Luanda to send the message that it was rooting out FLEC "terrorists," rather than peaceful activists.

The episode was a sharp reminder of why, despite their long list of grievances against Chevron, most Cabinda activists preserve the bulk of their anger for the Angolan state, which they accuse of widespread human-rights abuses, as well as a generally thuggish approach to the province's problems. In 2004 New York–based Human Rights Watch issued a briefing paper that included reports of gang rapes, torture, and sexual humiliation carried out by the Angolan Armed Forces. I wanted to hear some of these stories firsthand, so one morning, with the help of several activists from Mpalabanda's youth wing—and 50 kwanzas (about $0.60) to convince the guard to look the other way—I sneaked into Cabinda town prison to talk to three men who had been described to me as "political prisoners." Once inside, I heard the men's accounts of predawn raids on their homes and trumped-up accusations of "treason" and "terrorism." One was accused of plotting to bomb the airport and the other two of beating up an Angolan bishop. All had been held for three months without hearing a charge or appearing in court. All were members of Mpalabanda and believed that was the real reason they were there.

There is only so much time you can spend in Cabinda meeting with young men from a banned organization and being sneaked into prisons before your actions start to look suspicious. On my second day in town, therefore, I presented myself to the authorities and asked for an interview. I was told the governor was not available on such short notice, but one of the two vice governors might be free. Later that afternoon, I was summoned back to the provincial-government offices for what I was informed would be a very brief meeting with Antonio Goma, the vice governor for technical, economic, and community affairs.

Goma explained he had to meet a delegation at the airport in fifteen minutes, but promised to help however he could. Eager not to waste time, I sat down and asked him about the troubles in Cabinda. The nervous-looking press officer suddenly remembered that we had not been formally introduced. The three of us rose to our feet stiffly as she went through the formalities, and when we had sat back down, Goma proceeded to spend the next fourteen minutes welcoming me to Cabinda and telling me all about the province's geographic situation—land area, population, natural resources, livestock, and so on—in a manner that made it clear I was not to interrupt. When he had finished with this little verbal almanac, Goma said that Mpalabanda claimed to be a "civic association" but was actually a political movement, and therefore illegal under Angolan law (which allows political activity to be channeled only through organized national parties). He then glanced at his watch, and the press officer dutifully reminded him that he was running late. As Goma got up to leave, they both smiled flatly and said they hoped the meeting had been useful. "I didn't see you taking a lot of notes," the press officer added as she ushered me out of the building a few minutes later.

Luckily for me, Cabinda's other vice governor proved a little less inclined to hide behind formalities. João Mesquita, vice governor for social affairs, simply arrived unannounced at my hotel later that evening and offered to sit down with me over a beer on his way to an engagement. Mesquita rattled off a series of social and economic development programs he said were under way in Cabinda and insisted that the Angolan government did not believe in "the path of war" as a solution. "It is antagonism between brothers of the same nation," he said, knowing full well just how strongly the overwhelming majority of Cabindans would object to such a characterization of the conflict. Mesquita claimed that FLEC's "principal protagonists" had been reintegrated into society, and that the Angolan soldiers stationed in Cabinda (whose exact number he professed not to know) were there only to protect the province's timber, gold, and oil industries—he failed to say from what—and to perform a humanitarian function, providing "medicine and health care" and rebuilding bridges and roads. The army, he said, was "having a very positive impact on the population, especially the rural population."

It was hard to imagine a point of view more out of touch with that of the local population, or a version of events more likely to get Cabinda nationalist blood boiling. Sadly, though, it seemed entirely consistent with the arrogance I had been led to believe was characteristic of Angola's centrally appointed governors. After all, in 2003, according to its own balance sheet, the provincial government of Cabinda had seen no problem with spending $1.8 million on cars, $120,000 on mowing the governor's lawn, $85,000 on a Miss Cabinda contest, and $2.4 million on "Christmas gifts."

To hear the other side before leaving Cabinda, I tracked down one of the most outspoken independence activists—and, I quickly discovered, one of the most intense. Father Jorge Congo, a Catholic priest with a doctorate in theology from Rome, had been agitating against Chevron and the Angolan government since 1993, and the passage of years seemed only to have sharpened his rhetoric. Under normal circumstances, Father Congo would not have been hard to find; but in 2005, he was one of many priests in Cabinda who had taken the bizarre step of going on strike and withholding communion as a form of protest against the appointment of an Angolan bishop in the diocese. Over the door of his pink colonial church on the beach just outside town were the words *"Eu sou a porta"* (*I am the way*, or, more literally, *I am the door*). But the door itself was bolted shut.

When I found him in his rectory, great mounds of rice and fish stew were laid out on the red plastic tablecloth in front of him and he was swatting away the dozens of mosquitoes that had begun to circle. "Look around," he said almost immediately as the mosquitoes kept landing. "Is this a town that produces nearly a million barrels of oil a day?" Malaria is rife in this part of Africa, and it is never worth taking chances, so before long our interaction descended into the farcical spectacle of two grown men trying to talk while smacking themselves across the face or chest every few seconds.

But even the most aggressive of mosquitoes couldn't dull Father Congo's edge. "For thirty years, Angola has ruled us by force," he said, the bridge of his nose wrinkled with indignation. "They treat us like dogs in our own homes." I gently tried to put some of the vice governor's assertions to him and got as far as the suggestion that Mpalabanda was a political movement before he jumped in, enraged. "Is it politics to say we want clean water?" He was nearly shouting. "If we say we want to promote the culture and traditions of the Cabindan people, is that politics?" He smacked himself on the back of the head and then added, a little more quietly, "Look, as a priest, I don't want to see people dying. I want to be finished with war too. But with *dignity*."

◄◄─►►

Across Angola, whether the issue is Cabindan separatism or the disappearing billions queried by the IMF, or the fact that there has not been an election for fourteen years, the MPLA are quick to blame the war when their stewardship of the country is called into question. But as the months since the war ended in April 2002 turn into years, more and more critics around the world are losing patience with this tendency to play the "war card." The war is over, they say. You have been in power for over thirty years, and it's time to start seeing some results. But probably even more relevant than the civil war is what went before it. The legacy of extreme centralization and institutionalized resistance to independent scrutiny is one that the MPLA inherited directly from the Portuguese.

When the British and French ruled their respective chunks of Africa, they operated with at least a veneer of participatory democracy or, if nothing else, the fig leaf of economic protection. The Portuguese presence in Africa was a different matter. For its final few decades, it was the direct extension of the Fascist regime in Lisbon: one that, until its dying breath, saw Africans as little better than animals needing to be civilized. While British Prime Minister Harold Wilson talked of the "winds of change" blowing across the Empire in the early 1960s, and while Britain and France gradually prepared their colonies for independence, the Salazar regime operated under the "Statute of the Portuguese Natives of the Provinces of Angola, Mozambique and Guinea," an apartheid system that separated natives into savages and *assimilados*, who enjoyed most of the rights of Portuguese citizens. When independence came in 1975, it came as the accidental result of a socialist coup d'état in Lisbon that overthrew the Fascists. In a matter of days, 340,000 white settlers, who had managed the businesses in Angola and run the colonial bureaucracy, left Luanda. In Mozambique, São Tomé, and Guinea-Bissau, successful and united independence movements had emerged in the 1960s and quickly took power. But in Angola, rival liberation movements had been fighting each other for years, and when the Portuguese left, the new-born country plunged immediately into civil war.

Colonial Angola, like Fascist Portugal itself, had been characterized by repression, press censorship, and a total lack of official opposition. There was no tradition of multiparty politics or parliamentary democracy for the new Angolan rulers to use as a model, as there was for their counterparts in the former British and French colonies. It should not be surprising, then, that when the MPLA took power in 1975 and immediately had to go on a war footing against internal and external enemies, a highly centralized and institutionally paranoid top-down presidential system quickly took root in Angola. The MPLA's politburo approach to government was cemented by official exchanges with the Soviet bloc, and their paranoia made worse by a 1977 coup attempt that nearly toppled the government. The failed putsch, which grew out of a grassroots movement based in Luanda's *musseques*, had a profound effect on then-president Agostinho Neto and, almost ever since, the regime has maintained order using the dreaded paramilitary police units nicknamed "Ninjas." At no point in the country's history, whether at the Alvor Accords that gave Angola independence or at Bicesse, Lusaka, or Luena, have civil-society organizations or the Catholic Church been included in peacemaking negotiations. The only people given a voice throughout Angola's history have been those with guns.

Perhaps nowhere is this tortured legacy more visible than in Cabinda, where the only political discourse available to young activists is one of hatred and violence. When even a respected and intelligent Catholic priest in his forties can barely contain his rage during an interview, what hope is there for the province's youth, whose only organized activity is branded "terrorism" by the authorities? As I got up to leave after an hour-long meeting with Father Congo, he had a final warning for me to take away—one that, given the country's history, unsettled me deeply. For a moment, it even took my mind off the constellation of mosquito bites that, despite vigorous self-flagellation, had appeared up and down my arms and neck. "It is the Americans who are letting this happen," he said sadly. "The Cabindan people are starting to hate Americans." He paused and gathered himself, clearly trying hard not to sound too angry. "These people are just employees. They aren't responsible for the policies," he said, referring to Chevron staff. "But you see, people are starting to *hate* them."

CHAPTER 4

INSTANT EMIRATES

NIGERIA, ANGOLA, GABON, and Congo-Brazzaville, along with, to a lesser extent, Cameroon, make up an old boys' club of African oil exporters. Each has been on the map of sub-Saharan petro-states for the better part of a half-century now, and, until recently, the names of these five or six countries, along with those of Algeria and Libya to the north, were virtually synonymous with the words "African oil." A few other countries, such as Ivory Coast, Sudan, and South Africa, pumped a few thousand barrels of oil a day between them, but the other forty or fifty nations in Africa were largely ignored by the petroleum industry.

In the early 1990s, all that began to change, and change quickly. Four African countries—Equatorial Guinea, Chad, São Tomé and Príncipe, and Mauritania—have joined (or are about to join) the ranks of the world's oil-producing nations, and several more—including Mozambique, Madagascar, Uganda, Kenya, and Ethiopia—seem likely to follow. Another dozen appear more doubtful, but their leaders are crossing their fingers. All across Africa, it seems, every petty autocrat with a vested interest in distracting attention from his country's internal problems is holding licensing auctions and inviting the oil industry to gather seismic data.

Almost everyone you talk to in Africa, from the most senior politician to the humblest goatherd, is at least vaguely aware that there is an "oil boom" under way on the continent, and that somewhere, somehow, there is *seriously* big money to be made. But many are also painfully aware that Africa's history with oil exploration has been complicated, at best, and that seriously big money has usually brought serious misery for Africa's people. Each of the "original" members of the continent's petro-club has seen its particular experience with oil distilled into an aphoristic shorthand—a warning to those who would follow in its path.

Gabon is the golden child ruled by a self-interested French puppet who forgot to prepare his country for life after oil and has left it with a castrated economy.

Cameroon and Congo are much the same story, but in the latter country, oil has fueled a bloody civil war that has left the population traumatized and afraid to speak out against the country's high-level corruption.

Angola is the sleeping giant where billions of dollars have disappeared and where the government maintains a deep distrust of and distance from the international community.

And as for Nigeria, it is simply the doomsday scenario, an amalgamation of all the worst oil has to offer Africa: corruption, ethnic hatred, Dutch disease, and rentierism, organized crime, militant rebellion, hostage taking, and sabotage of industry activity, and a country held together tenuously by a political establishment whose leaders, in the words of a U.S. government think-tank, "are locked in a bad marriage that all dislike but none dare leave."

With role models like these, you can only wonder what the citizens of Africa's newest petro-states must be thinking. Many of these overnight emirates are small, deeply impoverished nations that lack even vocational-education systems, let alone the old boys' decades of experience with negotiating with European and American multinationals. In some cases, this lack of experience has made it possible for small, shady companies to swoop in and sign exploration deals that are deeply unfavorable to the host countries. In general, traditional political ties with fading colonial metropoles have taken a backseat to a rambunctious and unpredictable way of doing business that favors the more nimble and the more aggressive of the industry's operators, as well as plucky upstarts from regions of the world not known for their trading links with Africa.

For example, in Mauritania, a former French colony, the 33,000-barrel-a-day Chinguetti offshore field was brought onstream in 2006 not by the French multinational Total, but by Woodside Petroleum, a small Australian independent that has also picked up promising acreage in Uganda, a former British colony. Mauritania's onshore concessions have gone to Hardman Resources, another, even smaller Australian company, as well as to a Chinese concern. Chinese companies have also been active in southern Sudan, where a combination of civil war, U.S. sanctions, and NGO advocacy concerns have kept North American and European companies away for over two decades. The Chinese have been joined there by Indian, Malaysian, and Kuwaiti companies.

But one of the most striking features of the African oil boom as it began to take shape in the 1990s was its American flavor. For years, African oil was essentially a European story, featuring large European multinationals. Many of these companies had established lucrative beachheads in Africa during the colonial era and relied for some years afterward on social, educational, and cultural networks that linked the political elites of the host countries to those of the former European powers. For Britain and France, in particular, companies such as Shell, Elf, and, to a lesser extent, British Petroleum, were able to reap the rewards of this historical relationship. In the early years of the twentieth century, European oil companies were simply an extension of the sprawling colonial bureaucracy that ran Africa. In November 1938 Shell D'Arcy, the British exploration wing of the Royal Dutch/Shell Group of companies, was granted an exclusive Crown license to prospect for oil throughout what was then the Protectorate of Nigeria. In 1956 Shell made its first discovery, deep in the swampland of the Niger Delta, near the village of Oloibiri. After independence, Nigeria invited other companies as part of an effort to diversify its commercial and political relationships and reduce its dependence on Britain, but to this day Shell remains the biggest producer in Nigeria, with Shell's fields accounting for about half the country's output.

Meanwhile, under the auspices of Société des Pétroles d'Afrique Equatoriale Française (SPAEF), France's state-owned oil company Elf Aquitaine began drilling along the coastline of the lush tropical forests of Central Africa in the early 1950s, when what we now call Gabon, Congo, and Cameroon were still part of a vast and thinly populated rain forest known as "French Equatorial Africa." Elf was eventually privatized and merged with the Franco-Belgian company TotalFina to become TotalFinaElf, but not before an embarrassing scandal exposed its murky business activities in Africa as a sort of "secret arm" of France's warm relations with African dictators. The company, which has since rebranded itself as "Total," is not only one of the world's largest multinational oil companies, but also still one of the biggest players in Africa, with massive operations in Angola, Gabon, Congo, and Nigeria.

In the 1970s, however, though African oil looked likely to remain dominated by Elf and Shell, the American company Gulf Oil developed offshore fields in Angola. Mobil, along with the Italian giant Agip, became more active in Nigeria. Then, in the late 1980s and early 1990s, deepwater drilling techniques that had been perfected in the Gulf of Mexico began to become more commercially viable, although still something only the majors could really afford. The rapid availability of this technology, along with the buzz generated by a few unexpected finds by independents in the 1990s, was about to make African oil interesting again.

As a result, the African petroleum landscape today has a decidedly more American accent to it than it did a decade or two ago. ExxonMobil, once present only in Nigeria, today operates the massive Zafiro and Ceiba fields in the waters of Equatorial Guinea, and has become a major player in Angola, alongside its compatriot, Chevron. Exxon has even built a 600-mile pipeline to deliver crude from the parched semi-savannas of southern Chad to the coast of Cameroon. Smaller American companies, such as Marathon Oil and Amerada Hess, have also become heavily invested in the Gulf of Guinea.

But perhaps nowhere has this new American footprint been more visible than in Equatorial Guinea, one of the world's smallest, most bizarre, and most unlikely of countries.

◂◂▸▸

Very little about Equatorial Guinea can be described as "normal." A self-parodying burlesque of a tin-pot kleptocracy, it is ruled by a man who killed his uncle to become president and has been accused of everything from international money laundering to eating human testicles. The country is made up of three fairly unrelated pieces of real estate: a lone rectangle of jungle wedged between Cameroon and Gabon, a tiny volcanic island hundreds of miles away near the Nigerian coast (and home to the capital Malabo), and an even more far-flung and thinly populated island so unloved that Spain is allowed to use it as a toxic waste dump. Equatorial Guinea's history has been one of isolation and eccentricity at the best of times, laced with terrifying displays of savagery at the worst. For decades it was one of the most forgotten and impoverished countries on earth; but since the discovery of prolific quantities of oil and gas in its waters, it has received a lot more attention from the world. So much so that in 2004 a gang of South African mercenaries, financed by a Lebanese businessman and the son of a former British prime minister and flying in an American plane piloted by Armenians, tried to stage a coup there. Their goal? Replace the president with a former colleague whose rivalry with him had begun in the 1970s with a disagreement over a girl.

The peculiar trajectory of Equatorial Guinea's history can best be understood in the context of its status as the only former Spanish colony in sub-Saharan Africa. On a continent once dominated by Britain, France, and Portugal, such institutions or informal groupings as Francophonie or the British Commonwealth allow Africa's independent nations to maintain strong linguistic and cultural connections to others of the same colonial lineage. Moreover, because virtually all African countries function under educational, political, and legal systems similar to those bequeathed to them by the former colonial powers, a strong institutional divide exists between the Anglophone and Francophone blocs. It is a division reinforced by transport links: the path of least resistance for those wanting to travel between two neighboring African countries that speak different languages still, far too often, involves a flight to Europe. Scraped together from the territories of Spanish Guinea and given independence (almost by accident) in 1968, Equatorial Guinea has never been able to tap in fully to a larger community of African states, and has remained on the fringes of the continent's politics.

Perhaps more important, Equatorial Guinea's independence came when Spain was still an international pariah ruled by Generalissimo Francisco Franco and his fascist Falange party. For years, the only European political philosophy to which the territory's elites and civil servants had been exposed was authoritarianism. President Teodoro Obiang himself received his education at the military academy in Saragossa. Meanwhile, Spain, desperate to come in from the cold and gain international credibility, and keen to dispense with three useless scraps of African jungle it had inherited in a treaty with Portugal in 1778, jettisoned Guinea at a hastily convened independence ceremony. Franco, hoping to demonstrate his bona fides to the world, ensured that a free and fair election took place in Malabo, the only democratic election he ever organized. There was only one problem: The wrong man won.

Francisco Macias Nguema, a former court translator, a paranoid schizophrenic sociopath, and the uncle of Equatorial Guinea's current president, quickly declared himself "President for Life" and, for good measure, "Immaculate Apostle of Steel" and "Unique Miracle of Equatorial Guinea," among other titles. The cinematic depravity of his regime was rivaled only by the likes of Uganda's Idi Amin. In

1975, for example, Macias celebrated Christmas Day by lining up 150 of his political opponents in a soccer stadium and shooting them dead while a brass band played Mary Hopkin's 1960s anthem, "Those Were the Days, My Friend." On another occasion, thirty-five prisoners were told to dig a ditch and stand in it. The trench was then filled so that only the men's heads stood out of the ground. Within twenty-four hours, ants had slowly eaten the prisoners' heads, and only two men remained alive.

As a young man, Macias had failed the civil service exam three times, an experience that had left him with a hatred of the bourgeoisie that ran so deep that as president, he actually banned the word "intellectual." Schools were closed, hospitals were ordered to use only traditional African medicine, and the country was quickly emptied of a generation of doctors, teachers, lawyers, and journalists. Macias turned increasingly to marijuana and hallucinogenic drugs, and began making decisions based on conversations with imaginary advisers. Malabo's electricity supply was crippled after the president ordered employees at the city's only power plant to stop using lubricant. He would use "magic" to keep the machinery moving, he insisted, but the plant quickly exploded. When citizens tried to flee the country by water, Macias banned fishing and ordered all boats to be destroyed.

During the Macias years, anywhere from one-third to one-half of Equatorial Guinea's population either died or fled the country, as it became known in diplomatic circles as "Africa's concentration camp." The British and French media had their hands full of colorful African troublemakers—from Amin to Bokassa to Smith—and Spain, struggling with the transition to democracy following the drawn-out demise of General Franco, was in no position to draw the world's attention to an unraveling African disaster story. A small and strategically unimportant tropical backwater, the bastard child of a backward European power, Equatorial Guinea went completely ignored by the "civilized" world, and Francisco Macias, Immaculate Apostle of Steel, literally got away with murder.

That is, until August 3, 1979, when Macias was overthrown and executed in a military coup engineered by his nephew and right-hand man, Brigadier General Teodoro Obiang Nguema Mbasogo, a canny political survivalist who celebrated twenty-five years as president and "Liberator" in 2004 and shows no signs of wanting to step aside. As chief of internal security on Bioko Island throughout the 1970s, Obiang was more than privy to the savagery of the Macias regime. His own rule has lacked some of the most extreme excesses of his uncle's, but he has not hesitated to rely on mass detentions, intimidation, and torture to suppress opposition. In 1999 the U.S. State Department went public with documentary evidence detailing instructions given to security forces to urinate on prisoners, kick them in the ribs, slice off their ears, and smear oil over their naked bodies to attract stinging ants. According to one person who survived the ordeal, the other prisoners were slowly beaten to death on Obiang's personal orders. The State Department says Obiang's government employs "the psychological effects of arrest, along with the fear of beating and harassment, to intimidate opposition-party officials and members." One Western diplomat has called Obiang a "known murderer."

In the early 1990s, Spain began taking a more progressive view of international cooperation and, along with the United States, put pressure on Obiang to introduce at least the appearance of multiparty democracy. Obiang relented, but not before crushing the fledgling civil-society groups and opposition parties with a round of mass detentions. The subsequent poll in 1996 saw Obiang elected president with 99.2 percent of the vote. His extraordinary popularity had slipped a little by the time he faced reelection in 2003, when he received only 97 percent of the vote. It was a fairly strong finish for the opposition party, whose leaders languished in jails and torture chambers.

Spain's reward for its behind-the-scenes attempts to democratize Equatorial Guinea was the opportunity to sit back and watch a closer relationship develop between Malabo and Paris. In the 1990s Obiang opened the Ministry of Francophonie and began cultivating closer ties with neighboring Francophone states such as Gabon and Congo—to the point of adopting the Central African franc as Equatorial Guinea's currency and French as a sort of unofficial language of business—in a clear shot across the bows of Madrid.

And it was not only the Spanish who were snubbed. In the early 1990s, John Bennett, the American ambassador in Malabo, made a series of increasingly critical remarks about Obiang's poor human-rights record and the country's lack of genuine democracy and rule of law. In response, several U.S. buildings—including the American embassy and the Peace Corps office—received a letter, thrown out of a car window shortly after midnight one night, accusing Bennett of being a senior Klansman and suggesting that he be shot dead. Bennett's inquiries revealed the car to be owned by a regime loyalist who later went on to become minister of justice, and, eventually, prime minister. "You will go home as a corpse," the letter read at one point. In 1995, during a round of budget cuts, the State Department shuttered the embassy in Malabo.

In retrospect, it was probably the worst possible time for Washington to distance itself from Equatorial Guinea, as it was around then that a number of small independent American oil companies, like Walter Oil & Gas and Ocean Energy, began prospecting off the nation's coast and hit a jackpot. When seismic data and initial-exploration wells suggested in excess of 500 million barrels of oil and gas reserves in the country's territorial waters, some of America's bigger players moved in. Today ExxonMobil, Marathon, Amerada Hess, and others have between them invested $5 billion in bringing Equatorial Guinea's oil onstream, and the nation produces some 290,000 barrels a day—a hefty figure for a country of only 500,000 people, and a figure that, for a while, made Equatorial Guinea sub-Saharan Africa's third-largest producer, behind only Nigeria and Angola.

The corresponding transformation of Equatorial Guinea has been as dramatic as it has been swift. Fifteen years ago, Malabo had just one hotel, where electricity, food, and running water were never guaranteed. The telephone directory was two pages long and listed the lucky subscribers by their first names, along with their four-digit phone numbers. Today a dozen overpriced hotels are always booked up, a French company has built an efficient mobile-phone network, and a shiny, air-conditioned airport has replaced the old ramshackle hut of a reception hall. Off the record, diplomats once sneeringly referred to Equatorial Guinea as "the armpit of Africa"—a reference not just to its insalubrity but also to its geographical position as a small, damp swatch of jungle sitting under the raised "arm" of West Africa. Today, with direct air connections to Houston, Paris, and Amsterdam and one of the world's fastest-growing economies—at least on paper—Africa's armpit has become Africa's overnight emirate, and even boasts a new nickname: the "Kuwait of the tropics."

Yet, in a country so small and so oil-rich that it ships close to one $50 barrel of oil for each of its citizens every single day, this wealth has had little impact on the population. Despite a per capita income that, at $6,200, is one of the highest in Africa, the majority of Equatorial Guinea's uneducated, underfed, and disease-ridden citizens live in stinking, festering slums without running water or electricity, often cheek by jowl with opulent government ministries. Malaria is more rampant than almost anywhere else in Africa, life expectancy is just fifty-two years, and the majority of the population still struggles to get by on less than $1 a day.

Where has all the oil money gone? For clues, you could visit the exclusive Washington suburb of Potomac, Maryland, where Obiang owns a $2.6 million mansion with ten bathrooms, seven fireplaces, and an indoor pool, which he paid for entirely in cash. The president also bought a smaller home in Maryland for $1.15 million, and his brother owns a $349,000 home in Virginia. You could visit the Los Angeles enclave of Malibu, home to movie stars and Hollywood moguls, where in 2006 Obiang's son, Teodorín, bought a $35 million mansion to complement the $6 million Bel-Air home from which he runs his hip-hop label, TNO Records. Teodorín's love of L.A. is probably not unrelated to the occasional relationship he has maintained with the American rap star Eve, but initially it grew out of his time at Pepperdine University in the early 1990s. The tab for his education was picked up by a small Texas oil company that quickly grew frustrated by the inability of their $50,000 stipend to keep up with the demands of Teodorín's lifestyle. Since his college days, he has spent much of his time living in five-star hotels in Paris and Rio and collecting Lamborghinis and Bentleys. Several years ago, French television showed him sashaying through Paris in one of his Bentleys and buying thirty tailored suits within a few hours. The editor of the influential newsletter *Africa Energy Intelligence* called him "the nearest thing there is to an African oil sheik."

A recent South African court case revealed how, in 2004, Teodorín had descended on Cape Town and in the course of a weekend, dropped more than $1 million on three cars—two Bentleys and a Lamborghini Murcielago—as well as nearly $7 million on two luxury homes. He had ordered lavish appointments for the houses and had apparently asked for the bill to be sent to the government of Equatorial Guinea, though in the end, the contractor was never paid and the renovations did not go ahead. The cost report seen by the court, however, showed that for one house, he had ordered a $150,000 Bang & Olufsen entertainment system installed, as well as a $15,000 plasma-screen TV, with $13,000 speakers. As for the kitchen, it was to be fitted with a $6,000 Gaggenau fridge, and a wet bar complete with a $1,500 Scotsman ice machine. The young man told the court that "both houses were in need of extensive renovations" and protested that "they were not in a fit state to carry out any entertainment."

A true Renaissance man, Teodorín is not only an international playboy but also a civil servant. Miraculously, he finds time in his busy social calendar to serve his country as minister of forestry, where he has imposed crippling operating charges and taxes on the timber industry in order to line his pockets. However, some of his more extreme antics have been the source of significant heartache to his father, who had once groomed him as a successor, to the irritation of other members of the ruling family. In December 2003 dissatisfaction within the Nguema clan over Teodorín's mischief nearly resulted in a palace coup, and the gossip during my time in Malabo was that the young man had been sent to drug rehab in Cuba, following an incident in which he almost killed his own uncle. Obiang's younger son, Gabriel, has since risen in favor and been put in charge of the all-important energy ministry, with overall responsibility for the oil portfolio.

In fact, it is probably fair to say that Equatorial Guinea is less of a functioning country than it is a lucrative family business that happens to come with a flag, an anthem, an army, and a seat at the United Nations. Obiang's ruthless brother Armengol—whom one former Western diplomat described to me as an "illiterate monkey-hunter with a taste for SUVs"—runs the feared internal-security service. Another brother is head of the armed forces, and Obiang's brother-in-law is ambassador to the United States. Several uncles, cousins, and nephews also serve in government. In all, in 2006, twenty-one of the country's fifty cabinet ministers were direct relatives of the president.

◄◄►►

While the overwhelming majority of the country's output is produced by Exxon, Marathon, and Hess, even the smaller players in Equatorial Guinea—companies like Noble Energy, Devon Energy, and Triton oil—are American. Several thousand of their employees are flown in and out of Equatorial Guinea on twenty-eight–day shifts, alternating between weeks on the offshore rigs and weeks of what often amounts to a fairly testosterone-driven form of R&R in Malabo. The rig workers—more "oilfield trash"—are housed in giant walled compounds designed to provide the comforts of suburban America. Clubrooms, cinemas, tennis courts, swimming pools, and American television are all standard. All the food—down to the last tomato—is flown in from the United States, and residents can even telephone their loved ones without dialing an international prefix.

Outside the compounds, most of Malabo's residents live in squalid shantytowns. Twice a day, the women sweat under heavy buckets of water they must fetch from wells for their families, knowing nothing of the fabulous comforts that exist on the other side of the razor wire. The only contact many of them will ever have with *los expatriados* is when the rig workers venture out of their compounds in search of a pair of spread legs—just about the only luxury Exxon-Mobil doesn't fly in from America. Female visitors are strictly forbidden in the compounds, but personnel are not prevented from venturing into town, where pleasures of the flesh can be readily indulged in for not much more than the cost of a restaurant meal and taxi fare home for the girl in question. Informal prostitution, virtually unheard-of ten years ago, has become a cottage industry in Malabo, but far more common is the phenomenon of young women hoping to forge enough of a connection with an expatriate oil worker to get themselves out of Equatorial Guinea for good.

The teenage girls in strappy tops and shimmering miniskirts who congregate in bars and hotel lobbies around town aren't the only sign of Malabo's changing fortunes. A decade ago, this capital city, home to a few thousand people, was a sleepy grid of a couple dozen tiny streets, trimmed with crumbling, colonnaded Spanish buildings with pastel porticoes and wrought-iron balconies. Until 1998 none of the country's streets was even paved, and there were only a handful of cars on the entire island of Bioko, where Malabo is located. Today, hulking cargo trucks hauling great lengths of pipe burp and belch in the town's endless traffic jams. Black S-class Mercedeses with tinted windows speed members of "the Family" from engagement to dubious engagement. And Filipino laborers are shuttled around in buses bought in bulk from U.S. cities disposing of unwanted stock. One group of buses from the San Francisco Bay Area still announce their destinations: "Tam Valley," "Tiburon," and "Concord Plaza."

As of 2005, Equatorial Guinea had become the third-largest destination for U.S. investment in Africa, behind only Nigeria and South Africa—an extraordinary achievement for such a tiny country. And with so many Americans working, living, and investing in Malabo, it was only a matter of time before the United States reopened its embassy there, as it did in 2006. But the decision was condemned by human-rights groups and seen by many as a cynical move motivated by Equatorial Guinea's growing importance as a supplier of oil to the United States. "I have no doubt that the only reason there has been interest from the State Department is the oil," Frank Ruddy, who served as Ronald Reagan's ambassador in Malabo, told me when I met him in his office in Washington in late 2004. "Why else would we build an embassy there? How else do you explain it? Everyone knows [Obiang] is a thug and a gangster and we are giving him all this praise. That is just shocking to me." Whatever the American motivation, Obiang seized on the State Department's decision as a vindication of his rule, and celebrated the news as "an act of transcendental importance in the history of Equatorial Guinea."

And a hard-won victory, he might have added. Since 2003 Obiang's government had retained the services of several high-profile Washington lobbying firms to help break down resistance on Capitol Hill and in the executive branch to the possibility of a thaw in relations between the United States and Equatorial Guinea. C/R International, a lobbying and PR outfit, received a $300,000-a-year contract to deliver "daily contact and interaction at the highest level of the U.S. government, including the White House, State Department, the Pentagon, and Congress, to send clear and concise messages about the need [for] a new and positive approach in U.S. relations with Equatorial Guinea." C/R also promised to disseminate "the positive aspects of the reality in Equatorial Guinea for American and international public opinion."

C/R International is run by Robert Cabelly, a former U.S. diplomat who specializes in putting a kinder, gentler face on African governments that have run into trouble in Washington because of their human-rights records. In October 2005, for example, C/R International signed up the government of Sudan as a client, as Khartoum sought to fend off hostile campaigns waged by the Christian right and by human-rights groups. But Cabelly has also represented nearly every oil-rich government in Africa. Between 1996 and 2002, Cabelly billed the government of Angola $6 million as it sought to bring its civil war to an end and establish closer ties with its historic enemy, the United States. And in 1995, as the military dictatorship in Nigeria outraged the world by hanging Ken Saro-Wiwa and his fellow Ogoni activists, and the U.S. Congress prepared to impose an embargo on the country's oil exports, Cabelly received $1 million from a petroleum company owned by the Nigerian president's son. C/R, along with eight other Washington lobbying and PR firms, went to work, and the legislation passed against Nigeria was heavily watered down.

And Cabelly is not the only lobbyist the government of Equatorial Guinea has in its corner. In 2005 it hired the Republican lobbyists Barbour Griffith and Rogers for $37,500 a month, as well as the PR firm Annabel Hughes Communications. It has also retained the services of Washington heavyweight Cassidy & Associates, for a hefty fee of $1.4 million a year. Bruce McColm, who runs the Institute for Democratic Strategies (IDS), has been one of Obiang's most tireless champions in the United States. "Strengthening democratic institutions" is the stated mission of IDS, but over 90 percent of its budget is devoted to promoting Equatorial Guinea. McColm's group monitored the last round of municipal elections in the country and, with no apparent irony, declared them "effective and transparent."

Though precise cause and effect is always hard to prove when it comes to the results of lobbying and PR campaigns, there is good evidence that Obiang's extreme makeover has been convincing to officials in the Bush administration. In 2002 in New York, Obiang met President Bush at a meeting of African heads of state, and in 2004 he was invited to Washington for talks with then—Secretary of State Colin Powell and Energy Secretary Spencer Abraham. In April 2006 Powell's successor, Condoleezza Rice, again welcomed Obiang to Foggy Bottom, and kicked off the grip-and-grin photo op by referring to him as "a good friend" of the United States. And, despite human-rights objections raised by members of Congress, MPRI, a Virginia-based military contractor run by Pentagon retirees, was given the green light by the State Department to train Equatorial Guinea's coast guard in the protection of offshore oil installations.

While the Bush administration was quietly laying the groundwork for closer relations with Equatorial Guinea in the years following September 11, 2001, hoping a skeptical press and public would buy the story that the country was improving its ways, a bipartisan group of lawmakers on Capitol Hill had other ideas. In 2004, following a particularly damaging series of articles by Ken Silverstein in the *Los Angeles Times,* the Senate Permanent Subcommittee on Investigations launched a full-scale investigation into allegations of financial irregularity in the handling of Equatorial Guinea's accounts at Washington-based Riggs Bank. The report, which was released in March 2005, exposed a host of questionable arrangements.

The Senate investigation found that at any one time, as much as $700 million of Equatorial Guinea's treasury deposits were held at the Dupont Circle branch of Riggs Bank in Washington. By itself this is not a crime. In fact, Riggs, the oldest bank in Washington,* had long prided itself on the discreet services it provided international clients. (Riggs liked to call itself "the bank of presidents" and, in its final years, fully 20 percent of its revenue was coming from embassy accounts in Washington.) But it is unusual for a government to keep its national accounts in a foreign bank, and even more suspicious when,

---

*In 2005, Riggs was bought by PNC Bank and its nameplate retired.

as in the case of Equatorial Guinea, the president is one of only three people allowed to sign for that account. (The others were his son and his nephew.) Obiang has justified the arrangement to Western journalists by describing oil revenues as a "state secret" in Equatorial Guinea or by noting that "there is a lot of corruption in Africa" and arguing that by restricting access to the Riggs account, he was able to keep an eye on things and prevent improprieties from taking place.

But Obiang's personalized approach to due diligence was not the selfless act of patriotism he made it out to be. According to the Senate report, the $700 million Riggs account was only one of more than sixty accounts from Equatorial Guinea that the bank administered between 1995 and 2004—and at least half functioned as purely private accounts for Obiang, his wife and sons, their extended families, and senior government officials and their families. The bank's lax oversight allowed money to be transferred freely between private and state accounts, and even senior bank management seemed unconcerned by the large sums sloshing around their vaults. "Where is this money coming from?" a Riggs vice president wrote gleefully in an e-mail to his colleagues. "Oil—black gold—Texas tea!"

During this period, Equatorial Guinea became the bank's biggest single customer, which is saying something, considering that Riggs also handled accounts for Saudi Arabia and former Chilean dictator Augusto Pinochet. At one point, Riggs wired $35 million from the main Equatorial Guinea Treasury account into the accounts of "two unknown companies" with addresses in the Caribbean—no questions asked. The Senate report even details two rather colorful episodes in which the Riggs official in charge of the accounts walked the mile back to Dupont Circle from the Equatorial Guinea embassy, carrying suitcases stuffed with $3 million worth of $100 bills wrapped in plastic cling film.

American oil companies were not spared by the Senate report, which found that Chevron, ExxonMobil, Marathon, Hess, and others had paid $4 million to fund scholarships allowing children of Equatorial Guinea government officials to study at universities in the United States. Investigators also uncovered some unusual business and real-estate dealings in Malabo. The sprawling ExxonMobil compound where oil workers live, for example, was "rented" for $175,000 a month from a company managed by Obiang's wife. Amerada Hess, meanwhile, rented offices for $445,800 from a fourteen-year-old relative of the president. Most embarrassing of all were revelations that direct payments that U.S. oil companies had made into Equatorial Guinea government accounts at Riggs had been immediately transferred to offshore banks. Several of these companies were, at the time of writing, facing further investigations under America's stringent Foreign Corrupt Practices Act.

The corruption and criminality of Equatorial Guinea's ruling family has long been a running joke among the diplomatic community. In one famous anecdote, a top aide to the president was arrested in New York's John F. Kennedy Airport after federal agents noticed a trail of marijuana spilling from his suitcase as he wandered through the terminal. And for much of 2002, at least, as AOPIG and others began touting West African oil as an alternative to the Middle East, the Bush administration appeared to be counting on the country's relative obscurity and risible lack of professionalism to shield it from scrutiny. When the flurry of damning stories about Equatorial Guinea began to come out in the U.S. and international media, and the Senate subcommittee launched its investigation, some of Equatorial Guinea's biggest cheerleaders in Washington knew they had a problem on their hands.

As embarrassing stories in the international press about private jets and Maryland mansions began to multiply like rabbits, Washington's enthusiasm for Obiang cooled noticeably, and Malabo found itself more shunned than ever by the international community, especially Spain, whose upstart Conservative government had never warmed to

Obiang or enjoyed the close relations that the more established Social- ists had nurtured since the 1970s. In Malabo, as well as abroad, gossip about the regime's internal stability began to swirl, with Obiang's prostate cancer and the jet-setting lifestyle of his heir apparent seen as particular liabilities. At an energy conference held in London, dele- gates openly discussed rumors of a plot to overthrow the government of Equatorial Guinea.

On March 7, 2004, the worst-kept secret in African oil politics fi- nally became a reality—almost. A cargo plane carrying sixty-four mercenary soldiers from all over southern Africa was intercepted by Zimbabwean officials when it stopped for "refueling" at Harare air- port. The plane, a decrepit 727 once owned by the U.S. Air National Guard (and most recently sold by a small company based in Kansas), was being flown by an Armenian crew, but was registered to a South African company with an address in the British Virgin Islands. Zim- babwean authorities quickly discovered that the mercenaries had stopped in Harare on their way from South Africa to Malabo to pick up a $180,000 consignment of black-market military hardware. The next day, fifteen men were arrested in Equatorial Guinea. They were believed to be part of an advance party meant to meet the plane on its arrival. Thanks to a helping hand from his old friend Robert Mugabe, Obiang had narrowly avoided being overthrown in an internationally organized coup d'état.

For months afterward, chins wagged with elaborate theories about exactly what happened in March 2004, and the failed coup has since gone down as one of Africa's all-time classics: a farcical scheme put together by a bunch of aging white mercenaries and financed by some of the most underemployed and overprivileged members of the British establishment (TALES FROM CLOUD COUP-COUP LAND read the headline in the British daily *The Guardian*.) For his part, Obiang wasted no time in lashing out at Spain, Britain, and the United States, hinting darkly at their involvement. He even filed a lawsuit in London against British nationals he believed were directly implicated, demanding they be extradited to Equatorial Guinea.* Certainly there has been no shortage of reasons to believe that the three governments were not entirely unaware of the botched operation before it happened. Just days before the attempted coup took place, for example, Spain moved several warships from bases in Spanish waters off the Canary Islands to a location near Malabo. Meanwhile, a steady stream of exiled Equatorial Guineans had passed through Washington in late 2003, each of them holding meetings with State Department officials and hoping to pass himself off as the only credible alternative to Obiang.

One of these would-be presidents was Severo Moto, who was given political asylum by Spain in 1982 and from Madrid heads what he calls the "Government of Equatorial Guinea in Exile," a surprisingly slick operation complete with regular cabinet meetings, press releases, and a stylish Web site worthy of a European government. In November 2003, Moto came to Washington for a four-day program of meetings and press engagements put together by Joseph Sala, a former State Department official to whom Moto paid $40,000. Sala, who runs a lobbying outfit called the ANN Group, helped Moto navigate Washington and press his case as the rightful heir to any post-Obiang Equatorial Guinea.

Among other candidates who received a hearing was the hapless Gustavo Envela, a Los Angeles–based actor whose career highlights include a small role in the movie *Sgt. Bilko* and an appearance on *Wheel of Fortune*. (Envela had been to Equatorial Guinea only twice for a total of less than two weeks since his family fled in 1970.) Equatorial Guinea's own anemic official opposition party, the barely tolerated CPDS, also sent a representative, but was unable to send its leader, who remained in prison in Malabo. Of all Washington's suitors

---

*The suit was thrown out in October 2006 by the Court of Appeal.

during 2003, though, it was clear that Moto was the slickest and best organized of the bunch—the kind of man you could do business with.

Moto must have impressed more than just the Americans. Evidence suggests that the mercenaries planned to install him as the new president of Equatorial Guinea, in exchange for a stake in the country's oil industry. The entire operation had been timed to run with clockwork precision. At 2:30 A.M. on March 8, the plane loaded with mercenaries was to arrive in Malabo, where it would be met at the airport by the "advance party" of fifteen men already in Malabo. From there, the assembled gang would split into teams, with one team securing the airport and the others heading into town to take over key ministries. A member of the government whose loyalty had been bought would guide two teams to the presidential palace and the room where Obiang would be sleeping. The president, along with Armengol and Teodorín, would be seized and taken to the airport and flown to Spain. Severo Moto would then land in Malabo at 3 A.M., gather supporters he claimed to have in the military, and declare himself president on national television. From start to finish, the operation was to have taken thirty minutes.

With his early history of complicity with the Nguema family, Moto is probably more of an opportunist than an oppositionist. He served as director of television and radio programming during the Macias regime in the 1970s, when he would have been responsible for official propaganda, and even served in Obiang's ministry of information until 1982, when the two fell out over a woman they both laid claim to and Moto was forced into exile in Madrid. Since then, the men have maintained a bitter and highly personalized animosity toward each other. In 1996, from Angola, Moto tried unsuccessfully to launch a coup against Obiang's government and was sentenced in absentia to ninety-six years in prison. Self-dramatizing and utterly unable to go for long without the glare of publicity, Moto is most in his element when he is able to place himself in the middle of a media circus, and

preferably one that makes him out to be a victim of elaborate conspir-
acies involving Obiang's henchmen. In the summer of 2005 Moto disappeared for more than a month, before turning up in Croatia, where
he held a press conference to tell reporters that he believed the Spanish government had been complicit in a plot to kill him. There is a
large question mark over just how much support Moto would have
had among Equatorial Guinea's population had he taken over as president in 2004. The sad reality might be that, with no tradition of popular participation in politics, no one in Equatorial Guinea would have
cared one way or the other.

Moto's status as a political refugee in Spain would be seriously
jeopardized if any illegal activity came to light, so it is no surprise he
has vehemently denied involvement in the 2004 coup attempt. Spain
has also denied advance knowledge of the coup, describing its movement of warships down the West African coast as part of "routine
exercises." One Spanish official who spoke to me on condition of
anonymity told me that his government simply "neglected" to announce the troop movement, which he described as a "PR mistake."

But both Moto's denials and those of the Spanish government
stretch credibility. Records show that on March 4, Moto checked into
the Hotel Steigen Berger in the Canary Islands, a Spanish archipelago
off the west coast of Morocco, where he stayed until the afternoon of
March 7. He is then believed to have been transported in a small
turboprop to Mali, where he prepared to take what he thought would
be his long-awaited victory flight into Malabo later that night. Instead, when news of the arrests came, Moto was flown quickly back to
Spain. It is inconceivable that the Spanish government, which had
kept a vigilant eye on Moto's activities for over twenty years, was unaware of these movements.

But what has raised even more eyebrows than Moto's activities and
cast the spotlight firmly on Washington is a pair of meetings held in the
weeks leading up to the coup attempt. The first of the two meetings

took place in Washington in November 2003, on the sidelines of a conference put together by the International Peace Operations Association (IPOA), an influential lobbying group that represents the interests of America's so-called "private military companies." In a dinner address to the assembled delegates, Theresa Whelan, the Assistant Secretary of Defense for African Affairs (the Pentagon's highest-ranking official responsible for Africa), spoke in glowing terms about the role that private military specialists could play in advancing American interests, not just in Iraq, as they already were, but also in places like Africa, where the U.S. military "may not want to be very visible." Afterward, Whelan was approached by Greg Wales, a British security consultant with close ties to the mercenary community in Africa. Wales gave Whelan his business card and suggested she might be interested in discussing some "trouble that was brewing" in Equatorial Guinea. The two had a brief conversation, whose details are a matter of dispute.

The second meeting between Wales and Whelan was far more substantive, formal, and premeditated. And its timing was far more suspicious. It took place on February 18, 2004, just hours before the coup attempt against Obiang was originally scheduled to take place. In the end, the plane that was going to fly in the mercenaries broke down, and the coup had to be postponed.

Whatever the nature of Wales's discussions with the Pentagon, he was not speaking from an uninformed position. A few days before his initial encounter with Whelan at the IPOA dinner in November, Wales received about $8,000 from Simon Mann, his good friend and the coup's chief architect. (Mann, along with two others, would be arrested on the tarmac in Harare on March 7, trying to load weapons and ammunition onto the ill-fated 727, and is now serving a four-year sentence in a Zimbabwean prison for his part in the operation.) Wales received a further $35,000 from Mann in January. Wales has denied having any knowledge of the coup attempt, but hotel records show he stayed in a room down the hall from Severo Moto between March 4 and March 6, 2004, in the Canaries.

Wales's friendship with Mann, the scion of a wealthy brewing family and a sixth-generation Etonian, dates back many years. A former SAS officer with deep interests in African oil and mining, Mann is also one of the original founders of the notorious mercenary outfit Executive Outcomes, which has been behind covert operations in some of Africa's worst trouble spots over the years, from Sierra Leone to Angola to Zaire. In November 2003, around the time that Wales first spoke to Whelan about Equatorial Guinea and received the initial money transfer from Mann, Mann signed a deal paying Nick du Toit $2 million for "unspecified projects," according to the British newspaper *The Observer.* Du Toit, a former South African special-forces commando with links to the arms trade, was the leader of the "advance party" that was waiting for Mann and the planeload of mercenaries to arrive in Malabo on the morning of March 8.

Clearly, Wales was familiar with the activities of his friends, if not directly implicated in them. And it is hard to believe that his conversations with the Pentagon were anything other than an attempt to gauge the Americans' attitude toward a possible coup in Equatorial Guinea. Du Toit has testified in court that unspecified "higher-up politicians" in the United States were informed of the coup ahead of time and gave it their blessing. Spain had even assured Mann that it intended to recognize the postcoup government, according to Du Toit.*

But if Washington and Madrid had some difficult questions to answer, they were nothing compared to the web of intrigue that enveloped London in the latter months of 2004 and that was about to take in some of the most prominent—and most reviled—figures in the British political establishment. From his prison cell at Chikurubi Penitentiary in Harare, Simon Mann attempted to smuggle a message to his wife on several scraps of paper, but they were intercepted by

---

*Du Toit's sworn testimony was part of his trial in Malabo, criticized by international observers for numerous irregularities. He has since retracted those parts that seemed to implicate the Spanish and American governments.

South African intelligence and quickly became some of the most talked-about documents in Westminster. "Our situation is not very good and it is very URGENT," the message began. Mann's lawyers

> . . . get no reply from Smelly and Scratcher. Asked them to ring back after the Grand Prix race was over! This is not going well . . .
>
> It may be that getting us out comes down to a large splodge of wonga!* Of course investors did not think this would happen. . . . Do they think they can be part of something like this with only upside potential—no hardship or risk of this going wrong. Anyone and everyone in this is in it—good times or bad. Now it's bad times and everyone has to F-ing well pull their full weight.
>
> We need heavy influence of the sort that—. . . Smelly, Scratcher . . . David Hart—and it needs to be used heavily and now. Once we get into a real trial scenario we are f****d.

"Scratcher" was believed to be Sir Mark Thatcher, son of former British Prime Minister Margaret Thatcher, an airheaded bon vivant playboy and perennial punching bag of the British press. (The nickname had been given to him at Eton, where he had suffered from eczema.) Sir Mark was a good friend of Mann (they were neighbors in the wealthy Cape Town suburb of Constantia) and, it later turned out, had supplied the plane that flew Severo Moto to Mali from the Canaries.**

"Smelly" referred to Elie Calil, a multimillionaire British-Lebanese businessman who has since emerged as the coup's chief financial

---

*British slang. "Splodge" is an undefined quantity, like a "load"; "wonga" is money.
**Thatcher was eventually found guilty by a South African court of supplying equipment for the coup, but he escaped with only a $400,000 fine, after claiming that he was not aware what the plane he had hired was going to be used for.

backer. Calil, who was born in Nigeria and made much of his fortune trading Nigerian oil, has a long history of questionable activity in West Africa. In 2002 French police tried to arrest him in connection with bribes he had paid former Nigerian dictator Sani Abacha. One former U.S. diplomat told me Calil was "the kind of guy who would make you want to reach for your wallet as soon as you met him, just to make sure it was still there." During much of 2003, Calil cultivated a friendship with Severo Moto, and appeared to support the idea of installing him in Malabo, in exchange for a one-time payment of $16 million and the promise of becoming Equatorial Guinea's chief oil broker. Records suggest Calil chipped in $750,000 of his own money to support the coup. In late 2003, shortly before the attempt took place, Simon Mann signed a $5 million contract with a group of Lebanese investors for "mining, fishing, aviation, and commercial security projects in West Africa." According to the *Observer,* the deal was a front for the entry of Elie Calil.

Calil has close ties to the Thatcher family and Conservative Party circles, and for a while in 2004, it looked like his association with the coup plotters might take some high-profile scalps. The best-selling novelist and Tory peer Lord Jeffrey Archer, one of the most universally loathed characters in British politics, was implicated when it emerged that he had made two calls to Calil's cell phone on the day the plotters had met in South Africa to finalize arrangements for the coup, and that Simon Mann had received a bank transfer payment of £75,000 from one "JH Archer."

But Calil's connections were not limited to Tories. A close friend of Tony Blair, prominent Labour politician Peter Mandelson, whose chronically scandal-tainted international business dealings have forced him to resign from cabinet positions on no fewer than three occasions, was also implicated. In 2000 the first of these scandals had forced Mandelson to give up a Notting Hill house he had bought with funds allegedly siphoned from the government, and led him to rent a

£500,000 flat from Calil in the exclusive London neighborhood of Holland Park. According to a report written for South African investigators by someone close to the affair, Calil discussed the planned coup with Mandelson, who "assured him he would get no problems from the British government side" and invited Calil to come and see him again "if you need something done."

By late 2004 rumors about the British government's complicity in the attempted coup had begun to accumulate, with the press reporting that CIA and MI6 agents had told senior military figures in Equatorial Guinea that if the coup took place they should sit tight and not try to defend Obiang—that they would be "well looked after" in return for their cooperation. Smelling blood, the Conservatives' shadow foreign secretary, Michael Ancram, decided to put a very simple written question to Foreign Secretary Jack Straw, his counterpart on the government benches: "To ask the Secretary of State for Foreign and Commonwealth Affairs when the Government was first informed of the Equatorial Guinea attempted coup plot." When the time came for Straw to answer Ancram's question on the floor of the House of Commons, the Speaker called out the reference number for the written question, and Straw rose and said, "In late January 2004," and then sat back down as the House continued with its business. It was an extraordinary admission for a sitting Foreign Secretary to make, and even more bizarre for its lack of explanation or spin. In just four words, Straw had acknowledged that the British government was aware of an attempt to overthrow a sovereign government weeks before it took place and had, apparently, done nothing about it.

In the end, the most likely explanation for what happened in March 2004 is that a group of private financiers based in Britain and South Africa tried to depose the government and install a new president, who would reward them with a share of the country's oil wealth, and that, to one degree or another, the British, Spanish, and Americans had foreknowledge of the venture but did little to stop it, aware that stepping in to protect a tyrant like Obiang could have disastrous PR consequences back home. For Obiang, however, the experience was a bitter wake-up call, a stark illustration not only that external forces were conspiring to destabilize his regime—possibly with the help of his own brother—but also that his friends in Europe and America could not be counted on to rush to his defense. With the exception of his wife and two sons, a few regime loyalists, and possibly Robert Mugabe, there was no one Obiang could really trust anymore.

Not that Equatorial Guinea had ever been a place where "trust" carried much weight in ruling circles. Obiang himself had come to power by overthrowing and killing his uncle. He has so little confidence in the loyalty of his army generals that, since 1979, he has placed his personal security in the hands of a private regiment of Moroccan palace guards. During the remaining months of 2004 and for much of 2005, however, the country plunged into an extreme paranoia, as Obiang began to see coup plots behind every interaction, and became particularly suspicious about the movement of foreigners in and out of the country. State radio urged citizens to report any expatriates behaving suspiciously, and troops stationed on every street corner in Malabo quizzed foreigners about their movements while truckloads of soldiers patrolled the streets near the presidential palace.

It was against this backdrop that I tried to visit Malabo in February 2005, as the one-year anniversary of the botched coup attempt approached and Obiang's relationship with the West continued to deteriorate. I had been warned by nearly everyone I spoke to, including the U.S. press attaché in Cameroon, that I should expect not to be let into the country, and that if I was, my activities would be under constant surveillance and I would find it impossible to get any real work done. The U.S. Senate investigation was in full swing, with its report due out any day, making journalists less welcome than ever in Equatorial Guinea. A few months earlier, American freelancer Peter

Maass, on assignment for *Mother Jones* magazine, had been accused of espionage, driven to the airport, and kicked out of the country on the orders of the president. The previous summer, an Australian TV crew had been approached at their dinner table one night and told that if they did not go straight to the airport, "bad things" would happen to them. A team from the British daily *The Times* had been arrested for taking pictures and released only after the British honorary consul sweet-talked the government. A reporter for AFP (the only Western press agency with a permanent presence in Malabo) had been jailed for nine days and beaten after he reported rumors of a coup. The list went on and on.

Year after year, in fact, the Committee to Protect Journalists has ranked Equatorial Guinea as one of the world's "most censored" countries. In 2006 it was in fourth place, just behind North Korea, Burma, and Turkmenistan. All broadcast media in the country are state-owned, with the exception of a TV network owned by the president's son. State-run radio has compared the president to God and broadcasts songs warning citizens that they will be crushed if they speak against the regime. There are no newspapers in Equatorial Guinea, and no bookshops or newsagents where people can buy foreign publications. The only magazines available are filled with pictures of a beaming Obiang shaking hands with foreign dignitaries. In 2005 a Spanish television network broadcast a program about the Riggs Bank investigation, and the government reacted by confiscating every satellite dish in sight.

Equatorial Guinea's brutal regime has always had a lot to hide, but until recently, no one had cared. Before the discovery of oil, there had been no leathery Afrikaners trying to overthrow the government, because it simply wasn't worth the effort. Rumor has it that at one point in the 1980s, Equatorial Guinea even tried to sell itself to Cameroon for $1 million, but its offer had been refused. You literally couldn't give the place away. Suddenly, though, between the U.S. Senate and the foreign press, a lot more people seemed to be asking questions, and the government was in no mood to make life easy for them.

Still, it seemed a shame not to make a good-faith attempt to visit Malabo and give the place a fair shake. After all, thanks to Obiang's desire to make friends in the right places, Americans were the only nationals not required to obtain a visa for travel into Equatorial Guinea, so it was just a question of turning up and taking my chances at the airport's immigration checkpoint. To my surprise, I made it past the flustered guard by self-importantly presenting a letter in English from my publisher and affecting a manner of impatient civility.

Within an hour of landing, I was perched on a barstool, chatting in Spanish with a pert Chinese barmaid, who wore a tight T-shirt that read: "I may not be perfect but parts of me are FUCKING EXCEL-LENT." In the corner, a group of large and rowdy tattooed Scots were gawping at a big-screen TV showing British pop videos. Outside, through a cluster of thick, damp palm fronds, truckloads of soldiers carrying machine guns could be seen driving idly past. So far, I thought to myself, no surprises.

I was with Mick Hoyle, a self-proclaimed "bargie" gypsy from the North of England with a smooth-shaven, turnip-shaped head, what looked like a Maori epic of tattoos under great tufts of arm hair, and not a trace of a neck. Mick had gone to the Falklands in 1983 and been offered an earthmoving job before ending up as a security guard on the rigs in the North Sea off Scotland. From there, it was on to Gabon and then Equatorial Guinea, where, since 2000, he had been in charge of security at Malabo's port.

Mick had thrown back three beers in the first twenty minutes and was suggesting I sample the local nightlife. A barbecue—apparently a popular activity among "oilfield trash"—was starting up across town, so we got into Mick's red Suzuki Samurai and drove off, with Mick periodically pulling over to roll down his window and howl and roar like a pit bull at the locals. When he had first arrived in Malabo, everyone had called him *el lobo*, "the wolf," because he was so hairy, and he still enjoyed playing on the joke. He admitted, though, that it wasn't just his body hair that earned him the nickname. "I was also violent," he said with a smirk. "I still get violent."

Nothing in Malabo is ever more than a three-minute drive away, and we quickly arrived at the barbecue, which was being held in a gated compound of freshly carpeted Southern California-style town houses that, in the warm night air, had the feel of off-campus housing in Santa Barbara. The party was ostensibly to celebrate a local girl's birthday, and the lads had pulled out all the stops. Giant ice chests filled with cans of Heineken, Corona, and Miller Genuine Draft had been hauled in, and a laptop projected J-Lo videos onto the wall. A couple dozen thirty- and forty-something-year-old men with mullets and bushy mustaches stood round in NASCAR caps and faded jeans cracking jokes. The rather fussy and expensive-looking hot buffet that had been arranged through ExxonMobil catering services was looking unloved, but it was probably the only thing saving the scene from looking like a twenty-year reunion of *Animal House*.

And then the girls started turning up. One by one they trickled in, wearing tight, breast-hugging dresses. In high heels and a skimpy red sequined dress, the birthday girl—a dead ringer for a twenty-year-old Condoleezza Rice—looked straight out of a music-video shoot in Miami Beach. One of the forty-year-old men with a mullet appeared to be her boyfriend. "Like what you see?" asked a male voice with a thick Scottish accent, from behind me. "Twenty quid and she's yours for the night."

The man's name was Jonno, and despite the fact that we had been thrust together as partners in lechery, we settled quickly into a fairly respectable conversation. Only a few of the girls were actual prostitutes, I learned. Most were really "nice family girls" looking for a husband and the accompanying ticket out of Equatorial Guinea, but willing to settle for gifts and meals in the meantime. "Sadly, though, most of the lads here treat them like ordinary slappers," he said. "Some don't even give them anything the next morning—they just boot them out the door." The general pattern, according to Jonno, was that "you go out with them once or twice and then they start asking you for things." He had once come back to his place to discover that a girl's brother had ordered a wide-screen TV to his address. The unamused retailer was waiting for him to pick up the tab.

I was lucky that this barbecue was being held in town by contract staff, rather than in the ExxonMobil compound by one of the company's employees. The latter live in far more luxurious detached houses, but they aren't allowed to bring in female guests. "If this same party was being held at ExxonMobil," Jonno explained, "you'd go knowing that it would just be a bunch of men standing around drinking beer." If an Exxon employee wants to go into town and pick up a girl, he either has to pay for an expensive hotel room or go back to the girl's place, which is likely to be a hovel she shares with her parents. "Exxon are terrified about their image," Jonno said.

An ExxonMobil staffer himself, Jonno gave me his cell-phone number and offered to show me around the compound one night so I could see how nice the accommodation was, with its swimming pools and tennis courts. No foreign journalists have ever been inside the infamous compound, and I salivated at the chance. But surely Exxon would never allow it? "Oh no, it's fine," said Jonno. "You're a bloke, so it won't be a problem getting you in." Even though I was a journalist? "Oh, yeah. As long as you're not a woman." It seemed an odd oversight for a company so worried about its image.

I was beginning to dismiss the horror stories I had heard about Equatorial Guinea as the usual tall tales that accompany discussions of African dictatorships. Over the next few days, I set up discreet background chats with oil-company officials, Western diplomatic staff, and even a few Equatorial Guineans. It was virtually impossible to get anyone to talk on the record, and the few times I spoke to locals, there was a lot of nervousness and darting eyeballs. There is no civil society in Equatorial Guinea—no NGOs, no advocacy groups, no independent newspapers. Even the Catholic Church, often the only safe channel for criticism in Francophone Africa, is afraid to speak out forcefully here. During one conversation on the outskirts of town, I was bundled out of sight and the door pulled shut behind me when my interlocutor spotted the minister of agriculture's car in the distance. I peeked out of the window to see a black Cadillac with tinted windows rolling slowly past and pausing in front of the door. I didn't bother asking why the car had Pennsylvania license plates.

One afternoon I got myself invited to a pool party at the offices of one of the American oil companies, an enormous mansion nicknamed the "Parthenon" for its classical columns in front. Rumored to have been built and rented to the oil company by a wealthy Cameroonian banker, the sprawling compound was like a cross between a Scottsdale vacation home and one of Saddam's palaces. I quickly learned the difference between the thousands of "oilfield trash"—a rowdy bunch of tattooed, womanizing roughnecks who split their time between their offshore rigs and their walled compounds—and this much-smaller community of white-collar managerial types who mostly live full-time with their families in private houses in Malabo. Here there were no mullets or ice chests of MGD and Bud Light. Instead, PR executives and liaison officers sat round the kidney-shaped pool and nibbled on kebabs and avocado salad prepared by the Greek chef from the posh Bahia Hotel (owned by Armengol). "This is almost like real life, Natalie," squawked an American woman from her deck chair. "Sitting around the pool, drinking beer." And indeed, it did all look very much like life in the American suburbs, as long as you ignored the ostentatious marble columns, and as long as you didn't notice the coils of razor wire snaking their way around the high stucco walls of the compound. As long as you didn't think too hard about the goats and the dysentery and the untreated sewage that the neighbors called "real life."

This turned out to be as far as I would get in Malabo. I didn't get ahold of Jonno again or see the fabled ExxonMobil compound before my visit to Equatorial Guinea was cut unexpectedly—but perhaps predictably—short.

⤙⤚

It had all started innocently enough. I had arrived in Malabo on a Friday evening, and, desperate to be seen doing everything exactly by the book, gone to the Ministry of Information first thing Monday morning to request press credentials.

Over the next three days, I paid five visits to the ministry, and a few other government offices besides, always, it seemed, lacking a particular form or passport photo or explanation of my purpose there. Everything was very friendly and relaxed, however. Wednesday morning, I was told we were just waiting for the minister's signature and I should come back at 3:00 P.M.

At the appointed time, I sat in the ministry waiting-room, admiring the peeling paint and the bare lightbulbs, and counting the drops of sweat trickling down my back. A fat, inebriated secretary with a sweet baby-face kept me company, flirting aggressively and asking me again and again if I was married or had children. "There are a *lot* of women in America," she said three times, each time stressing the word *"muchas"* more forcefully than the last, before uncorking a loud, boozy burp in my direction. After about an hour, she said it would probably be a few more minutes before my credentials were ready, and offered to buy me a drink. She led me out to a dirt courtyard behind the ministry building where chickens and goats wandered in and out of sheet-metal shacks. As we walked, people teased her about what a nice couple we made. We ducked into a stucco hut and sat on a sweat-dampened sofa in front of a television showing English Premiership soccer, while I sipped an orange Fanta and she drank a twenty-ounce bottle of beer.

Back at the ministry, a man who never introduced himself summoned me gruffly into an unlit concrete office with bars on the windows. He barked that my credentials were ready and all that remained was for me to pay the fee. I thought this might be an invitation for a bribe, but he pointed to an official-looking list of fees for various credentials, at the bottom of which was print journalism—and the sum of 300,000 CFA francs, or around $600. Fees for press credentials in Africa, where they exist (and they rarely exist), are generally under $50.

"*Oh, muchisimo,*" I said lightheartedly, having learned that good humor often worked wonders with stone-faced officialdom in Africa. But before me was a gatekeeper in no mood for joking around. He flew into a fairly convincing rendition of blind rage, shouting that 300,000 CFA was nothing to an American and that I was surely making good money writing about Equatorial Guinea. Remembering that the Americans he usually saw were oilfield trash wandering about town with nubile locals caressing their beer bellies, I tried to explain that I was working with a publisher's modest advance and that the money would be coming from my own pocket.

My Spanish wasn't up to the task, and he insisted there was no reason my publisher couldn't send the money straightaway. "Call them! Call them right now and tell them to send the money!" he shouted, thrusting his cell phone in front of my nose. "*Call them!*"

I looked at my watch. It was 10:00 A.M. in New York. I took the phone and started dialing the number. My editor seemed unimpressed by what must have sounded like a crude attempt to fish more money out of her. "Look, we can't start going down this road, John," she said. I explained that I wasn't actually asking for money, just following orders. She found someone in the office who could speak Spanish and put him on the phone to explain to the functionary how publishing contracts work.

"They say you are not even employed by them," he snorted when he finally put down the phone. "Clearly you can't pay the fee, so I can't let you do your work here." That seemed fair enough and I was about to tell him I would find a way to come up with the money, but he was no longer listening. If I wasn't going to pay the fee, then I would have to produce a sworn testimonial that I would never write about the country after I left. He whipped out a sheet of blank paper and slapped it down in front of me. *"Write it!"* he screamed at the top of his lungs as I fumbled for a pen. "Write it," he repeated much more quietly—almost in a whisper—"or else bad things will start happening to you."

I asked him if he wouldn't mind dictating the affidavit so there would be no risk of it being unsuitable, and dutifully took down the words he put into my mouth. "We'll be taking this on to the police station for notarization," he said as I signed and dated my statement. He then spotted the spiral notebook in my hands and it seemed to remind him of a rather amateurish oversight on his part. "You've been here for several days already, haven't you?" He snatched away the notepad before I could reply. "We'll need to see who you've been talking to while you've been here," he snapped, and then added, more calmly, "We can't have you going off and writing all the usual nonsense about how there's no human rights in Equatorial Guinea."

I began to apologize for the misunderstanding, and to assure him I would not be causing any more trouble, but he interrupted to say that he wanted to make one thing very clear. "I am not kicking you out of the country. I am not forcing you to leave." He was obviously tired of reading stories about "expelled" foreign journalists. "But you must understand," he added, "if you decide to stay, we can no longer be responsible for your safety."

It was time to leave. As I got up to walk out, he said he was now going to call the minister and tell him everything that had happened. I apologized again and slipped out the door as he picked up his cell phone.

I had lasted five days. A respectable run, I thought.

From the ministry, I went to my hotel, threw my belongings into my suitcase, and headed for the airport. The only flight out of the country was a $2,200 KLM red-eye to Amsterdam. The next Air Gabon connection to Libreville wasn't for another two days. I went back to my $120-a-night hotel feeling a bit stupid about the original $600.

I spent an uneasy night in the hotel room, slipping out to the Spanish cultural center across the street for dinner. But I began to think the incident had just been a piece of theatricality designed to put the frighteners on me and wondered whether I might try a couple of discreet interviews with oil-company officials in the forty-eight hours I had left in Malabo. It would have been a disastrous move.

Early the next morning, I was awakened by a call from Mick. "There's an e-mail going round," I heard him say, as I rubbed the sleep out of my eyes. "It's not good news."

Apparently, a photograph of me, along with the biography I had up on my Web site, had been sent to everyone at ExxonMobil and their subcontractors. The subject line read "new journo in town," and at the bottom of the bio, where I had written that I was "currently working on a book about African oil," the word "... allegedly" had been tacked on humorously, followed by the information that I was "dodgy-looking" and that I had "been asked to leave the country, tried to get on a flight last night, and last seen cowering in his hotel room."

The e-mail had been sent by Keith Brown, a British national I had not met but a close friend of David Shaw, an unctuous Brit who has worked as a consultant in the Ministry of Mines and Energy for many years. Shaw and I had met for drinks a few nights earlier, and he had spent a couple of hours trying to convince me that the Obiang regime got unfair press. He had even invited me to his office the next day and printed out some useful information about the country's oil industry. The guy had a lot of time for me, which should have been a tip-off.

That, plus he reported directly to the vice minister, Obiang's son Gabriel.*

As Mick read me the e-mail, I listened to the bit in my biography about a story I had worked on for *Newsweek* that had put me "on the sharp end of South African mercenaries" and smacked my forehead as the realization set in. In the realm of a brutal and paranoid dictatorship that didn't trust foreign journalists and regularly violated their rights, that had recently seen off a coup attempt by South African mercenaries, and that depended for its survival on a symbiotic and possibly corrupt relationship with the oil industry, there was now an e-mail going around ExxonMobil saying I had been told to leave and connecting my activities with those of South African mercenaries.

Looking back, my Armenian-sounding surname and Lebanese looks were probably not much of an asset either, given the provenance of some of those involved in the coup attempt.

Mick explained quietly and slowly that if I tried to talk to anyone at any oil company before I left, their government-liaison officer would be obliged to report the conversation immediately to the Ministry of Mines and Energy. I would be liable for arrest.

◄◄-►►

Back in Washington, John Bennett, the former U.S. ambassador who had been threatened with death in the early 1990s, laughed hard at my story. It was a frigid December morning and we had met over nachos and soft drinks at a Mexican café in the shadow of the U.S. Capitol building. Bennett, a burly teddy bear of a man with a neat white beard, has never shied away from speaking his mind when it comes to the Obiang regime's record on human rights. In 2004 he famously told *60 Minutes* that "if you've ever seen a man limp on both legs, you know you're in Equatorial Guinea."

Bennett painted a vivid picture of how much things had changed

---

*Remarkably, in 2006, Shaw became Britain's Honorary Consul in Equatorial Guinea.

in Equatorial Guinea with the arrival of the oil boom. In 1991, when he was first posted to Malabo, the State Department's idea of an "official car" for the ambassador was a 1984 Oldsmobile whose cloth ceiling had begun to sag from the humidity. Despite its age and poor condition, the car had fewer than 1,000 miles on its odometer—there were no roads to drive on in Equatorial Guinea. Moreover, only thirty-five Americans were in the country—mostly missionaries and a few employees of Walter Oil & Gas—and he had them all to his house for Thanksgiving. Today between 3,000 and 5,000 Americans are in Equatorial Guinea.

Bennett recalled a relationship with the authorities that was "so tense that if I was summoned by the foreign minister, I would kiss my wife good-bye beforehand, not knowing whether I was coming back home or going straight to the airport." By late 1993 relations had deteriorated to the point that Bennett was publicly accused by the government of performing witchcraft in an effort to alter the results of the country's legislative elections. The next day, he received a bemused cable from the State Department saying, "We're glad to see you're taking such an interest in the elections."

From that low point, relations between the United States and Equatorial Guinea could only improve, and improve they did when the oil began flowing. However, the Senate investigation into the Riggs Bank scandal, as well as suggestions that the United States was aware of the coup attempt of 2004, have made Obiang far more suspicious of Washington's intentions, and generally far less willing to align himself with Western powers. In October 2005 Obiang offered the United States aid for victims of Hurricane Katrina, but he also traveled to Beijing, where he held talks with Hu Jintao about Chinese participation in the Equatorial Guinea oil industry. On his return, Obiang declared triumphantly, "From now on, China will be our principal partner for the development of Equatorial Guinea," an announcement that sent a wave of terror through the American, Spanish, and French chanceries.

This shift of power from Europe and America to Asia is a microcosm of African oil politics in the 1990s and the first decade of the twenty-first century. It was a story I saw in Equatorial Guinea, in Angola, and again and again during the two short years I spent working on this book. Just as European majors had to make room for their upstart American counterparts in the 1990s, American companies are increasingly having to contend with newer arrivals to Africa's oil patch in the form of nimble, independent operators from Australia, Ireland, or Moldova, as well as the heavyweight national oil companies of China, Malaysia, Korea, and India. By any measure, the second great scramble for Africa is on, and shows no signs of slowing down. And the United States, bogged down in imperial adventures and nation-building exercises in the Middle East, is struggling to keep up.

In fact, many in Washington feel the United States is passing up a golden opportunity to engage more meaningfully with Africa's newest oil emirates, arguing that the miserable experience of Equatorial Guinea should not be allowed to become the norm. There are even genuine reasons for optimism, some claim, if we are willing to look for them; genuine reasons to believe, even, that oil can be a blessing rather than a curse in Africa.

This may well be true, and in the next two chapters we will look at attempts to put oil wealth to good use in Africa. But for the people of Equatorial Guinea, no amount of foreign goodwill will make a difference in the present climate. As long as the oil continues to flow and global prices remain high, and as long as the Nguema clan clings to power and treats Equatorial Guinea as an untouchable satrapy of personal wealth and privilege, few expect the people of this wretched country to do anything more than survive from day to day. Young men, if they are lucky, will be caparisoned in fatigues and issued guns and perhaps occasionally paid a salary, while young women will face the choice between long days spent sweltering under the weight of buckets of water and short nights spent panting under the weight of oilfield trash.

Nor is it likely anyone in the outside world will hear or read much about the daily struggle that is life in Equatorial Guinea. Since my aborted visit in February 2005, no Western journalist has successfully entered the country, with the wire services reporting on only the occasional plane crash or sporting event, and mostly from the serenity of Madrid or Libreville. The stories you have read here may be frivolous and yeasty, but they are also, for the time being, the last ones out of Malabo.

CHAPTER 5

PARADISE FOUND?

I COULDN'T WAIT to try the chocolate.

Before leaving home, I had seen a BBC documentary about Claudio Corallo, a man so obsessed with finding the perfect cocoa bean that he spent hours at a time bushwhacking through the thick jungle of his cacao plantation on the tiny and sparsely inhabited volcanic island of Príncipe, in the Gulf of Guinea. I had watched in amusement as the exasperated BBC reporter, drenched in sweat, struggled to keep up with his subject. The two cracked open pod after pod of cacao, none of which was ever quite good enough for Corallo.

São Tomé and Príncipe, a pair of tiny volcanic islands, is the second-smallest country in Africa. Until Portuguese explorers arrived around 1470, the islands were totally uninhabited and, even with today's population of 160,000, there are more palm trees than people in this almost-impossibly-beautiful corner of the globe. But its fertile volcanic soils happen to produce some of the best coffee and cocoa beans in the world. For centuries, the Portuguese brought slaves from the African mainland to harvest the extraordinary bounty of the islands. They built magnificent plantation estates, or *roças*, many of which functioned as miniature towns, complete with hospitals, schools, and chapels. Some of the *roças* even boasted their own railways.

But then, in 1975, the Portuguese left, and they left abruptly, just as they had in Angola. The *roças* were abandoned and quickly swallowed up by the islands' fast-growing tropical rain forest. Today, crumbling, peeling, and covered in vines, São Tomé's legendary plantations have become an exquisitely romantic sight, standing as silent, haunting monuments to a once-glorious colonial past.

Initially, the country's Marxist government tried to nationalize the *roças*, but the experiment was an abject failure. When São Tomé abandoned Marxism, the country began to seek foreign investors who might turn the plantations into sustainable cacao farms or boutique tourist destinations. And so it was that in the mid-1990s, Claudio and Bettina Corallo, a quiet Italian couple who for years had managed a cacao plantation in Zaire, fled that country's civil war, came to São Tomé, and bought the Nova Moca Plantation. Today the Corallos employ a handful of São Toméans at their plantation on Príncipe and in the one-room "factory" behind their home in the middle of São Tomé town, where they and their children pitch in to make a few dozen bars of chocolate a day. From start to end, everything is done by hand, and the finished product arrives at Europe's finest chocolatiers after a long journey in special heatproof containers.

When I met Bettina Corallo in São Tomé, Claudio was away on Príncipe, no doubt dangling like a lemur from the branches of a cacao tree, believing himself closer than ever to finding the Holy Grail of chocolate.

"Come and have a look at how we make our chocolate," she said, as she led me to a small shed in the back garden where the cocoa beans were being dried and roasted. Inside, she plucked three fresh pale-green beans from a batch that had come in a few days earlier. "Go on," she said, "try one." I hesitated, remembering from experience that raw cocoa beans have possibly the foulest and most bitter flavor of any fruit on earth. "These are different, you'll see." Bettina smirked like someone about to carry off a cruel practical joke. I popped one in my mouth obediently and chewed.

"They taste like olives, don't they?" And she was right. "That's what makes them so special," she said. "It's very rare for cocoa beans to be edible in their raw form, let alone taste good. These are very unique beans." I popped the other two in my mouth and, as we walked out, pocketed a handful for later.

Bettina took me to an air-conditioned room behind the house, where chocolate in a small vat the size of a beer keg was being melted down while a tray of the stuff cooled and was broken into small pieces by a pair of local women. A third woman was wrapping the final product in cellophane and ribbon and gluing simple handmade labels to the packages. Bettina handed me a chunk of the finished chocolate and stood back, beaming like a mother at her child's school play. Her hands were clasped in anticipation of the reaction she had seen so many times before.

I took my first bite and nearly wet myself. Smooth and smoky with a grainy, almost gritty texture, the stuff didn't even really taste like chocolate. Light and brittle, with just enough sugar added to bring out the intense flavor, it vaporized in my mouth. It was like being given a glass of vintage wine after a lifetime of drinking grape juice.

I forgot about Bettina for a moment, but then I looked over to see her waiting. I silently cursed the wall of broken French that stood between us. *"C'est incroyable,"* I finally managed. She looked genuinely relieved, then sent me on my way with a little bag of chocolate. Even in São Tomé, 100 grammes of the Corallos' chocolate sells for a hefty five dollars at the small (and only) shop in town. Fortnum & Mason of Piccadilly, suppliers to Buckingham Palace, "purveyors of fine foods" for nearly three hundred years, and one of the few places in the world to stock the Corallos' chocolate, sells 40 grammes of the stuff for eighteen dollars. The department store's chief chocolate buyer has described Nova Moca chocolate as "among the best in the world."

But I will admit that after several weeks on the wrong side of the African oil story—listening to the rants of armed militants and the veiled threats of police-state apparatchiks, and enduring miles of rutted roads and swampy creeks in the company of slippery politicians and tattooed roughnecks—anything might have tasted good. São Tomé and Príncipe is only three hundred miles from Bioko, the island on which Malabo is situated, but culturally, politically, and socially, São Tomé and Equatorial Guinea might as well be on different planets. One is a brutal, paranoid dictatorship that terrorizes its own population, while the other is a relatively open multiparty parliamentary democracy with no history of bloody coups or violent insurrections. One is a country run by a greedy ruling family that accumulates wealth and power at the expense of other, resentful ethnic groups, while the other is a country where all citizens are the descendants of Portuguese slaves and therefore share a common history of oppression and a conspicuous lack of ethnic hatred. One government assumes that every foreigner who turns up uninvited is plotting a coup, while the other promotes ecotourism and small-scale cacao farming. One nation is swimming in oil wealth, while the other is dependent on international aid for its survival.

However, the two countries do have one thing in common: geology. São Tomé, Príncipe, and Bioko are links in a chain of ancient volcanoes that stretches to Mount Cameroon on the mainland. And since the Gulf of Guinea became known as a heavyweight oil province, there has been considerable interest in these tiny islands. Analysts have spoken of a billion barrels or more in potential reserves, and speculators have quietly moved in to stake claims to offshore licenses. In August 2000 a poorly defined maritime border with Nigeria was settled hurriedly in anticipation of oil exploration, and a Joint Development Zone established by the two states.

But some analysts remained skeptical, suggesting São Tomé's potential as an oil producer had been hyped excessively by Arab-bashing neocons in the United States. Such skeptics point to the fact that in 2002, as the American press fêted São Tomé as the "new Saudi Arabia," an exploratory well had yet to be drilled, and the country was at least ten years away from becoming an oil producer even if there *was* any oil. And, to some extent, the naysayers have been vindicated by the international oil industry's lukewarm response to the exploration blocks made available by São Tomé over the past two to three years. Even in the most promising block of all, Chevron drilled a well in early 2006 and found the results disappointing. By early 2007 it is likely the company will have exited São Tomé altogether. Nevertheless, there is no question that São Tomé has some oil, and possibly enough of it to make the island nation a significant producer in coming years. And, perhaps more important, it seems to lack much of the baggage that can make oil companies nervous about investing in Africa.

In fact, one of the first observations many visitors make upon arriving in São Tomé is that it doesn't "feel like Africa" at all. There is no African food, no African music, and virtually no trace of traditional African religion. The clothes, the architecture, the cuisine are all European. All names are Portuguese and, with the exception of a dying island patois, Portuguese is the only language spoken. Those who have traveled to both Africa and the Caribbean are struck by just how much closer São Tomé appears to the latter. The place is a quintessential tropical island paradise. Groups of naked children play in the ocean, while teenagers wander along deserted beaches shaking coconuts out of palm trees. From the water's edge, rain forests climb up precipitous peaks that seem perpetually encircled by mist. Abandoned plantation estates, pink and crumbling, hide between mountains like forbidden cities. In town, pairs of boys speed around on rickety motor-scooters, and a handful of banged-up old taxis sit idly waiting for fares—but in one of the world's smallest and sleepiest capital cities, where it seems half the buildings are government ministries, it's often easier to walk. With its languid beaches and European lifestyle, São Tomé seems far more "New World" than African. Barbados without the tourists.

It is precisely this lack of "African-ness"—this conspicuous absence of ethnic conflict, instability, and government brutality—that both the international community and the oil industry are counting on. Development experts from around the world have taken a keen interest in São Tomé, eager to play their part in what might just turn out to be an African success story. And São Tomé has reciprocated. Its president, Fradique de Menezes, a darling of the West, has made all the right noises about wanting to ensure that the opportunity provided by oil wealth is not wasted, that revenue is handled transparently, and that oil does not become a source of conflict or economic stagnation. "I have promised my people that we will avoid what some call 'the Dutch disease,' or 'the crude awakening,' or 'the curse of oil,'" he told a rapt audience in Washington in 2003, with a beaming Colin Powell listening approvingly.

However, this puts the cart miles before the horse. Not only does no one really know how much oil São Tomé has, but São Tomé can be said, without much exaggeration, to be totally unprepared for life as an oil-producing country. The country suffers from endemic poverty and an extraordinary lack of what development experts call "institutional capacity."

Just how poor and underdeveloped is São Tomé? Not only is there no university in the country, but the one high school is so strapped for cash that it must teach São Tomé's students in three five-hour shifts, beginning early in the morning and lasting until late into the night. The country's national budget in recent years has averaged a mere $50 million, much of it coming from traditional crops such as coffee and cacao, or from fishing. An additional $35 million in the form of international development aid pours in every year, making the country one of the world's biggest recipients of direct aid as a percentage of GDP. The lion's share of this aid has traditionally come from Portugal and from Taiwan, which views its special relationship with São Tomé as a fairly inexpensive vote in its favor at the United Nations. (Unfortunately, the strategy has largely backfired, since São Tomé has lately proved unable to pay its $17,000 dues to the United Nations and has therefore been ineligible to vote in the General Assembly.)

The country is so poor, in fact, that it has resorted to some fairly creative fund-raising schemes. The São Toméan postal service once issued commemorative Marilyn Monroe postage stamps it hoped would turn into collector's items, supplying a substantial percentage of the nation's income. In recent years, one of São Tomé's biggest industries after cacao has become routing phone-sex numbers that have been banned in Europe and America through its telephone exchanges. In the 1980s, as part of a deal with the Spanish government, São Tomé even agreed to accept Basque political prisoners from France in exchange for an increase in foreign aid, in effect allowing itself to become a penal colony for ETA militants. Very few were sent over in the end, but a wizened and long-bearded man with leathery skin can still be seen hanging around in São Tomé's bars every evening, watching the sun set.

Possibly the only characteristic of São Toméan society more persistent and more predictable than its poverty and its lack of capacity has been the cliquishness and cronyism of its tiny political class: a few dozen Portuguese-educated technocrats, none of whom knows the first thing about petroleum geology or licensing rounds or international-exploration contracts. São Tomé politics has always been a claustrophobic affair—the exclusive preserve of an entrenched (and largely *mestiço*) elite. *"Somos todos primos"* goes the unofficial slogan of São Toméan politics—"We are all family here"—a cheerful boast about the country's lack of ethnic hatred that has gradually taken on an ironic significance in the face of endless corruption scandals. Since 1991, when the country gave up Marxism-Leninism for multiparty democracy, electoral politics has been a ferocious, fast-paced series of spats and feuds and shifting coalitions and partisan realignments, most simply formalizing personal or financial disagreements of the country's political leaders. Just since 1991, São Tomé has experienced fourteen changes of government—more than many African countries put together. The current president alone has gone through eight prime ministers since he was first elected in 2001. And in 2003 dissatisfaction with the way the country's future oil wealth was being managed helped provoke a coup that briefly ousted the president before the Nigerians stepped in and had him reinstated—reinforcing suspicions about who was really running the country.

Given this entrenched atmosphere of cronyism and handshake politics, set against a backdrop of endemic poverty, underdevelopment, and illiteracy, it is perhaps no surprise that São Tomé was party to one of the more bizarre and scandalous backroom deals the world of African politics has witnessed—one in which large chunks of the country's oil wealth was given away to an obscure group of Texan and Nigerian speculators with no experience in offshore-oil exploration.

◂◂-▸▸

Like many African states, São Tomé is not a complete stranger to oil exploration. As early as 1973, when the islands were still part of Portugal's African colonies, a license was given to the British firm Ball & Collins, who, along with the Texas Pacific Oil Company, drilled on the islands—only to hit a duster. In the early 1990s, the shady South African property speculator Chris Hellinger drilled a few more on-shore wells. They turned up some shale and heavy sands, but nothing that could be described as "commercially viable." Until the mid-1990s, it never occurred to anyone to try drilling offshore.

Then came 1995 and news of ExxonMobil's giant discovery in the deepwater Zafiro field offshore Equatorial Guinea. Existing seismic data suggested that São Tomé's waters were unlikely to contain the kinds of reserves being found offshore Equatorial Guinea, but a few enterprising explorationists noticed a poorly defined maritime boundary area between São Tomé and Nigeria that appeared to share many of the geological characteristics that had made Equatorial Guinea into a world-class producer. In early 1997 the government of São Tomé was approached by a small Texas concern calling itself Environmental Remediation Holding Company (ERHC), run at the time by veteran Louisiana wildcatter Sam Bass, Jr. Bass had an offer he felt the São Toméans would be hard-pressed to refuse.

There was a good chance the country was sitting on a jackpot, but one it was in no position to exploit on its own. ERHC would therefore pay the government of São Tomé and Príncipe $5 million for the right to negotiate on its behalf with other foreign oil companies interested in any future licenses that became available. ERHC would also market São Tomé as a destination for petroleum-exploration activity, in effect assuming the function of a state oil company, and acting as a broker for the country's oil licensing. In exchange for this "service," the company would receive preemptive rights on all future oil blocks, in addition to a raft of concessionary privileges. ERHC's small office suite in a suburban Houston business park would become the unofficial address for the São Toméan oil industry.

News of the deal was received with bewilderment and disbelief by oil-industry analysts, one of whom called it "a raid on São Tomé's future national treasury." ERHC, it quickly emerged, had started life in 1986 as the Colorado company Regional Air Group Corporation, and then morphed into an environmental-cleanup specialist, before transforming itself in 1996 into a independent exploration company focused on the Gulf of Mexico. It had, as far as anyone could tell, one full-time employee, no drilling equipment, and only $1.5 million in cash. It was unclear what, if anything, it knew about oil exploration.

As bizarre as the 1997 deal with ERHC was, it was only the beginning of São Tomé's problems. In July 1998 ERHC and São Tomé created a company called STPetro, which would function as São Tomé's national oil company. The government of São Tomé would retain a 51 percent stake, with the balance going to ERHC. In August STPetro and ExxonMobil signed a deal for a technical-assistance program. Since neither ERHC nor the São Toméans were equipped to undertake even basic seismic surveying, ExxonMobil would evaluate the hydrocarbon potential of São Tomé's offshore waters, in exchange for preferential rights to several future oil blocks.

Suddenly things started to unravel. São Tomé fell out with ERHC and its new CEO Geoffrey Tirman over a number of contractual terms. When Tirman visited São Tomé, the government accused him publicly of refusing to pay the full $5 million owed them, and Tirman responded by hastily arranging a press conference at which he claimed the country's prime minister, Carlos Gomes, had demanded bribes from him. Gomes threatened to arrest Tirman and charge him with sedition, at which point Tirman made a beeline for the airport.

The case went to arbitration in Paris and ended only when Tirman agreed to sell a controlling interest in ERHC to Chrome Energy, a Nigerian company owned by prominent Nigerian businessman Sir Emeka Offor, who has close ties to President Olusegun Obasanjo. As part of the settlement Chrome/ERHC was forced to renegotiate its contract with São Tomé, but retained an extraordinary number of privileges and concessions, including an automatic 15 percent stake in up to four exploration blocks and 10 percent of all São Tomé's future oil profits. According to one fairly conservative IMF scenario, Chrome/ERHC could easily reap $1.4 billion over the lifetime of the fields from its upfront investment of just $5 million.

Possibly the most interesting aspect of the 2001 settlement is that it was made conditional on a resolution of São Tomé's maritime-border dispute with Nigeria, strongly suggesting high-level Nigerian involvement. São Tomé's president at the time, Miguel Trovoada, was well connected in Nigerian government and business circles and it is possible that, when the São Toméan government fell out with ERHC, Trovoada approached Chrome's chairman for help. In February 2001, just before ERHC was sold to Chrome, the long-running dispute between Nigeria and São Tomé (which had begun as a disagreement over fishing rights) was brought to an end with the creation of a Joint Development Zone (JDZ) on terms extremely favorable to Nigeria. Despite São Tomé's far-stronger claim to the waters, Nigeria would receive 60 percent and São Tomé 40 percent of revenues from the two countries' exploitation of the JDZ.

Many in São Tomé felt the country had been taken for a ride—first by smooth-talking Texan wildcatters and then by São Toméan elites whose primary loyalties lay with their Nigerian business interests. President Trovoada's second—and final—term in office ended in September 2001, and a wealthy cacao farmer and relative unknown, Fradique de Menezes, was voted into office, in what some observers interpreted as an end to the era of questionable Nigerian interference. De Menezes, a short, round, and avuncular figure with an erratic manner and chummy sense of humor, was quickly embraced by Washington's oil evangelists in the months after 9/11. Glowing profiles of him appeared in the American press, most of which referred to him by his first name, and painted him not only as a reformist figure committed to transparency and good governance but also as a big cuddly teddy bear, fond of cracking jokes and giving hugs. "Fradique" paid several visits to Washington in 2002 and 2003, including one in which he was noted to have made a particularly positive impression on President George W. Bush.

De Menezes was happy to be portrayed as a break from the past, a beleaguered president trying to make the best of a bad situation, but it became apparent before long that he was not the white knight who was to rescue São Tomé. In February 2002 it emerged that Chrome had transferred $100,000 into the Belgian bank account of CGI, de Menezes's private company. De Menezes dismissed the payment as a "campaign contribution," but it only confirmed many people's impression that the new president—who is Belgian-educated and a former Portuguese citizen, and has no real support base in São Tomé—was a puppet of the hated Trovoada family. It didn't help matters when Emeka Offor's cousin Nnamdi Noruke admitted to the Nigerian magazine *Newswatch*, "Offor assisted the party of the former president to win their elections and when they won they installed the current president," adding that the Trovoadas had assisted de Menezes "financially and otherwise."

Soon after his election, de Menezes wisely attempted to distance himself from the Trovoadas, and in May 2005, finally sacked the former president's son, Patrice Trovoada (so loathed that he is the only person who travels round São Tomé with armed bodyguards), from his job as presidential oil adviser, after a round of bickering between Nigeria and São Tomé over the allocation of oil blocks. But de Menezes continues to be seen as a man too close to Nigeria for the tastes of São Toméans, most of whom feel a more instinctive kinship with the country's traditional regional ally, Angola.

◄◄—►►

For São Tomé, the ERHC episode has been a harsh reminder of how ill-prepared the country is for the complexities of managing a petroleum endowment in the face of powerful and experienced neighbors like Nigeria. But it has also been an excellent demonstration of how much the African oil boom of the past decade owes its existence to the work of "independents"—small, nimble operators that have made it their business to slip in under the radar and clinch deals that often turn out to be worth millions to their investors and the multinational oil companies that follow.

When the first stirrings of an oil bonanza were felt in Equatorial Guinea in 1991, it was not ExxonMobil or Shell or BP that turned up with their drills and platforms. It was a tiny Texas outfit called Walter Oil & Gas Corporation, which pumped 7,500 barrels a day of mostly gas condensate out of the country's territorial waters—peanuts in the context of global oil, but a lot for a company with only a dozen employees. And when the former French colony Congo was being written off as a "declining" producer in 1999, it wasn't the French multinational Total that decided to revive some of the marginal wells it had plugged up and abandoned. It was a publicity-shy French company called Maurel & Prom, which had been kicking around as a small shipping concern since the nineteenth century, that snapped up the acreage.

The definition of an "independent" company in the context of the oil industry is open to debate. Originally, in the United States in the early twentieth century, the term was used to describe any oil company that was not part of the Standard Oil group of companies— later called Esso and ultimately Exxon. This definition is long obsolete, though it is a piece of trivia that comes in handy when you meet a oilman at a cocktail party.

Broadly speaking, today, petroleum exploration is undertaken by three different types of companies: national oil companies (NOCs); large, integrated companies (commonly known as "majors"); and independents. National oil companies are either wholly or partly owned by a national government. They can be titans of the global oil scene, like Saudi Aramco or Brazil's Petrobras, but they can also be obscure and negligible extensions of the state bureaucracy, like PetroVietnam or Rompetrol of Romania. Although there is a growing trend for NOCs to be partially privatized or to compete for exploration licenses in other countries, most were originally intended as a way to keep a nation's oil wealth in the hands of its citizens and generate jobs and technical expertise. Since virtually every country has at one time or another made some effort to prospect for oil, there are almost as many NOCs today as there are flags at the UN—the great exceptions being in Europe, where almost all have been privatized and sold off.

The majors, on the other hand, are far fewer in number—a couple dozen, by most people's estimates. Unlike the NOCs, the majors are wholly private, generally publicly traded, and often multinational in structure. But like the NOCs, they are fully, or "vertically," integrated. In other words, they take part in every aspect of the petroleum business, from exploration, development, and production (the so-called "upstream") to transportation, refining, marketing, and sales at the pump ("downstream"). The majors are generally globally recognized brands, and a few of the larger ones, such as Total, BP,

Shell, ExxonMobil, and Chevron, have earned the nickname "super-majors" for their size and scale. ExxonMobil, the largest of these, has an annual turnover well in excess of the GDP of many of the world's poorer countries.

According to one definition, the independents are "everyone else"—all the companies that are neither NOCs nor fully integrated majors. The Independent Petroleum Association of America defines an independent oil company as one that makes its money "at the wellhead." In other words, one that operates in the world of exploration and production and doesn't get involved in the downstream process. By that definition, an independent is an oil company without a refinery or a filling station. Outside the United States, however, the category is a little more fluid. For example, the Irish company Maxol calls itself an independent even though it operates a network of filling stations. In global terms, it is probably safe to say that an independent can be any company that is small and nimble and operates largely out of the public eye, its activities followed only by industry analysts. An independent might well be publicly traded, for example, but generally not on one of the world's flagship bourses, like the Dow Jones Industrial Average or the FTSE. In fact, a simple rule of thumb might be that the independents are the companies you've never heard of.

Independents also tend to be essentially national, even provincial or parochial, in character. Indeed, the stereotypical independent in the United States is a company based somewhere in Oklahoma, run by retired geologists and petroleum engineers, and lacking the slick publicity materials, flashy Web sites, and overstaffed HR departments of the majors. An independent's expertise and core competence supposedly lies in exploration, particularly in small, experimental fields the majors consider too marginal or too risky. Drilling exploratory and developmental wells, independents hope to stumble upon a big find and then sell or "farm out" the acreage to bigger companies with the capital and technology to operate the field, and simply get rich off the profits. A company with this kind of approach to the business is often termed a "wildcatter," conjuring images of whiskey-swilling Texans and fly-by-night operations, though many companies calling themselves "independent" insist they often stick around for at least the production phase of a project, even if they don't get involved in building refineries or running gas stations.

Whether you see independents as bottom-feeders or pioneers, however, it is hard to deny that the business model they operate under was made for Africa, where risk and volatility are an accepted part of life, and where trust and personal relationships count for far more than glossy brochures and the size of your market capitalization. And indeed, while it might be irresponsible to reduce the story of African oil into a tale of swashbuckling robber barons from all corners of the earth descending on a tropical Golconda, there is no question that the weak negotiating position of many African governments has resulted in a number of distinctly dubious deals being signed. As the ERHC debacle in São Tomé demonstrates, there is a less-savory side to the independents' success story in Africa, particularly in those countries that have never before attracted attention from the petroleum industry. Only a few African countries, such as Nigeria and Angola, can boast the experience, technical expertise, and negotiating savvy required to manage an oil windfall. In some of the smallest and most impoverished countries on the continent, such as São Tomé, only a handful of people have even gone to college. This is not to suggest that independent oil companies set out to deceive, or to do questionable deals. The majority of independents are legitimate businesses with at least a passing concern for their reputations. However, a total lack of education and capacity in some host countries has made it impossible for government officials to distinguish between serious business propositions and the promises of snake-oil salesmen brandishing uncapped pens over dotted lines.

In the worst cases, this has led to some embarrassing, deceptive, and downright bizarre deals. The São Tomé–ERHC arrangement is only the most high-profile of these, thanks to the fact that São Tomé seems to have genuine potential as an oil producer.

However, there have been others.

In 2002, for example, the volatile West African nation of Guinea (not to be confused with Equatorial Guinea) made the extraordinary decision to award the rights to its offshore exploration and production—*in their entirety*—to HyperDynamics, a small Houston software company that had turned itself, seemingly overnight, into a frontier explorationist focused on Louisiana and Mississippi—and now, apparently, Guinea. Generally, sovereign authorities divide their territory into blocks and auction off the licenses to a variety of companies, but Guinea, for reasons unclear, put all its eggs into the basket of an inexperienced "minnow" (as independents are often called). Perhaps unsurprisingly, HyperDynamics never attracted a larger farm-in partner to develop its Guinea concession—but it still requested a drilling permit from the Guinean government in June 2005. A month later, the government abruptly canceled its production-sharing agreement, in a letter that was never received by the company. HyperDynamics first heard the news in the press. In language not typically associated with corporate executives, CEO Ken Watts blasted the decision as the result of "misinformation, lies, and deceit that had been propagated over the last three years by people we trusted implicitly."

But the prize for sheer audacity must surely go to White Nile, a company cobbled together by former England cricketer Phil Edmonds and Andrew Groves, a thirty-six-year-old investor whose father once worked for South African intelligence. Five months before the twenty-one-year north-south civil war in Sudan came to an end in January 2005, White Nile signed a deal with the leadership of the Sudan People's Liberation Movement for the rights to a 26,000-square-mile drilling concession in SPLM-controlled south Sudan. At first sight, making a splashy deal with a rebel movement in a war-torn African country might seem par for the course for Edmonds, a colorful character whose past business dealings have involved everything from salmon fisheries to luxury hotels to platinum mines in Africa. What was extraordinary about the SPLM deal, though, was that it was for a large and potentially lucrative license *already claimed* by the French multinational Total, which had abandoned exploration work when the civil war broke out in 1983. Shares in White Nile spiked thirteenfold overnight on London's Alternative Investment Market before trading was suspended pending investigation by the authorities.

Occasionally, independent oil companies operating in Africa are not just ballsy but downright dubious. Energem, a Canadian independent that operates in more than a dozen African countries, started life as DiamondWorks before changing its name in June 2004. As rebranding decisions go, it was something of a no-brainer. DiamondWorks, a mining company with controversial business interests in the civil wars of Sierra Leone and Angola, was originally created out of a merger with a company controlled by Tony Buckingham, a British businessman with close ties to the notorious Executive Outcomes and other members of the mercenary community in Africa. Today Energem's chief executive is the Formula One magnate Tony Teixeira, a Portuguese–South African who was accused by the British government of gunrunning and selling fuel to UNITA rebels in Angola (allegations he has denied).

Sometimes it seems that Africa is filled with colorful chancers like Tony Buckingham, many of them refugees from the "white redoubt" that once fought so fiercely for Rhodesia and apartheid South Africa, now left casting about for a quick buck and a moment of relived glory. Tales of their antics are swapped over cold beers on languid nights as part of a never-ending blood sport among journalists based in Africa. But the serious point that these desperadoes help to illustrate is that there is still serious money to be made in Africa if you know how to make the right connections and play your cards right.

How else to explain the rags-to-riches story of VAALCO Energy, a company with just twenty employees that in 2002 was making a mere $445,000 a year in profits from its dwindling wells in the Philippines, but that has since turned its Etame concession offshore Gabon into a $34 million-a-year cash cow? Or what of the breathtaking success of Maurel & Prom, a small family company started in 1813 by two prominent Bordeaux shipping magnates? For most of its existence, Maurel & Prom specialized in shipping goods to the French colonies in West Africa and producing groundnut oil from its factory in Senegal. Its owners probably knew less about drilling for crude oil than the average first-year geology undergraduate. As maritime shipping declined in the 1970s, the company turned its attention to food processing, farming chicken and fish. In 1995 the state water and electricity company in Madagascar bought out what was left of Maurel & Prom and used its assets to develop mining, forestry, and shipping interests. The arrangement came to an end in 1999. Maurel & Prom might have muddled on like this for a few more decades had it not had the good fortune to be in the right place at the right time when the Republic of Congo was selling its Kouakouala oil exploration block that year. In 2001 the company built a feeder pipeline from Kouakouala to the M'Boundi field, where it had made a second, much-larger discovery. Today Maurel & Prom pumps over 11,000 barrels a day from Congo, and the return on its initial investment had reached 212 percent by the first half of 2005. In real terms, it had turned a trickle of seed money into millions of dollars in revenue. The company has since dumped its "non-oil assets," for which it says there is "no room on the balance sheet."

In the years since São Tomé emerged as a future oil producer, Chrome has tried to clean up its image, replacing its CEO with a former Marathon Oil exec and adding the former U.S. ambassador to Nigeria to its board of directors. But this has not erased the question marks over the future of São Tomé and Príncipe as a reliable place for the oil industry to do business. During the turbulent presidency of Fradique de Menezes, São Tomé's image has been damaged by a seemingly endless flotilla of corruption scandals. In 2004, for example, it emerged that several senior politicians had embezzled thousands of dollars from the country's official international-aid fund, which is intended to provide humanitarian relief for the country's poorest citizens. President de Menezes sacked his prime minister, who in turn blasted the "macabre and hideous masquerade organized by the President to demonstrate his greed for totalitarian power," and demanded that de Menezes explain the $100,000 he had received from Chrome Energy. She also pointed out that de Menezes's private company, CGI, appeared on the list of companies owing money to the general-aid fund. Smaller but no-less-significant corruption scandals included an illegal telecom deal secretly struck between the government and a Greek company registered in the Virgin Islands, and the April 2004 flap that saw the director of the country's only hospital fired following revelations of extreme mismanagement, embezzlement, theft of drugs, and the failure to keep even a basic balance sheet registering expenditures and receipts. From 2002 to 2004, 80 percent of payments collected by the hospital went unrecorded.

As one embarrassing revelation after another came to light during the first term of President de Menezes, and talk grew of oil wealth on the horizon, the political temperature in the country began to rise, particularly during the first months of 2003. Living conditions of the population seemed as bad as ever, but people connected to the government were increasingly spotted driving flashy imported cars. On April 11 eighty prominent citizens signed an open letter to the president, expressing concern over the country's economic hardships as well as the lack of transparency in negotiations with international oil companies. Less than a week later, an angry demonstration by market traders turned into a violent attack on a government office. The police opened fire and shot one man dead—a first in the country's history.

Finally, on the morning of July 16, 2003, while de Menezes was in Nigeria, a group of mercenaries seized control of government ministries, banks, the national broadcast center, and the airport. The coup was, from start to finish, a quintessentially São Toméan affair. Government officials were detained in a comfortable air-conditioned room at the army barracks and allowed to use their cell phones and receive meals brought by their families. The prime minister, who suffered from high blood pressure, was admitted to the hospital, where she was guarded by soldiers but received visitors freely. A few shots had been heard in the morning, but there were no casualties. The city remained calm and markets and shops stayed open. One week later, it was all over, with the putschists forgiven and allowed to go back about their business.

The coup had been staged by sixteen former members of the infamous Buffalo Battalion—a splinter group of Angolan rebels who had been trained and equipped by South African defense forces during the 1970s and 1980s. The two-thousand-strong battalion, which had fought alongside UNITA in the Angolan civil war, included fifty-three São Toméans who had left São Tomé as right-wing exiles from the Marxist government that controlled the country from 1975 to 1991. Trained in Namibia, the São Toméans had been given South African citizenship by the apartheid regime as thanks for their services. The Buffaloes, who have close ties with the notorious South African mercenary outfit Executive Outcomes, were also involved in plotting the failed 2004 coup in neighboring Equatorial Guinea.

Before launching their rebellion, the sixteen former Buffaloes approached Major Fernando Pereira of the São Toméan army, who for years had been complaining publicly to no avail about the poor living conditions of his soldiers. In a June 15 letter to the president and prime minister, Pereira had pointed out that the army barracks lacked water, working toilets, and basic medicines, and soldiers had to struggle on a monthly salary of $10, while government ministers were seen giving their secretaries and children new cars. Sensing a potential ally, the Buffaloes informed Major Pereira that they were plotting a coup and that they were prepared to fight the army if it got in the way. The ex-mercenaries were held in awe on São Tomé for their years of combat experience in Angola, Congo, and Sierra Leone. The São Tomé military, a ragtag Dad's Army of two hundred part-time fishermen, was no match for even sixteen Buffaloes—and everyone knew it. Pereira agreed to cooperate on the condition that the coup be bloodless and efficient.

◄◄─►►

"Sixteen?" He laughed hard as he licked bits of steak from between his teeth and looked around at his friends, who were smirking like schoolboys. "We could have taken control of São Tomé with just seven of us!"

I was talking to Arlecio Costa, one of the leaders of the 2003 coup, in the hushed, air-conditioned comfort of the Marlin Beach Hotel near the bumpy landing strip that passes for a national airport. It was an appropriate choice of venue. The Marlin Beach is owned by Chris Hellinger, a South African businessman who made a fortune in Angolan diamonds and always appeared to be happy to work with both sides during the Cold War, despite his known connections to South African intelligence. As Costa ran down a litany of complaints about São Tomé's governing elites, in what can only be described as a surprisingly charming South African–inflected Portuguese accent, I interrupted gently to remind him that in most countries, if you overthrow the government, you don't find yourself lounging about in luxury hotels afterward, entertaining foreign journalists. He shrugged sheepishly and conceded the point. "But look, we never intended to actually take power," he added with a smile. "We just wanted to shake things up a bit."

It is unclear whether they accomplished that much, however. "Today, everyone can see it," Costa continued. "People don't trust politicians anymore. There is instability everywhere. Just look around you. It is everywhere!" As a bow-tied waiter delivered another round of beers to our table, I looked out of the window at the beach and the swaying palm trees, and down at the half-eaten steak in front of Costa. At least for that moment, I had to admit I found it hard to see the instability.

Though the laid-back character of the 2003 coup speaks volumes about the nature of São Toméan politics (during a coup attempt in 1995, the rebels' tank ran out of fuel halfway to the presidential palace), even more revealing is the way in which the episode was brought to an end. President de Menezes was at a conference in the Nigerian capital Abuja when the coup took place, and it was Nigeria's president, Olusegun Obasanjo, who phoned Arlecio Costa to demand an explanation. According to de Menezes, Obasanjo told Costa that if he insisted on being "irrational," the Nigerians were also capable of being irrational. In the days that followed, a team of international mediators brokered a deal with the rebels and, on the evening of July 23, de Menezes was deposited in São Tomé on board the Nigerian presidential jet, in the company of Obasanjo. Two other Nigerian planes carried a regiment of Obasanjo's presidential guard, an entourage of Nigerian officials, and Nigerian journalists poised to capture their man in his full stride as champion of constitutional democracy in Africa. The two leaders went together to the presidential palace and then on to a signing ceremony in front of the UN offices, where de Menezes was officially restored as president with Obasanjo looking on. It was a spectacle that did very little to endear de Menezes to a population already concerned about the extent of Nigerian interference in the country's affairs.

In fact, behind virtually every debacle—from the coup to the corruption scandals to the dodgy deals with ERHC and Chrome—the invisible hand of Nigerian business interests has been seen or suspected. This speaks to São Toméans' highly emotional and deep-seated fear of Nigerians, based in part on the fact that Nigeria's more politically sophisticated population is nearly 1,000 times that of the islands. Moreover, Nigeria has a well-deserved reputation as the nerve center of organized crime for the region. In recent years, brash Nigerian traders have transformed the once-languid market in the middle of São Tomé town, often arriving with boatloads of counterfeit goods under cover of night, when the São Tomé coast guard (which consists of fifty men and two inflatable dinghies) is powerless to stop them. The Nigerians, who are far more experienced with the hustle of petty trading and shun the Portuguese habit of three-hour siestas, have literally pushed many São Toméan merchants from their stalls and onto the streets outside the market. Locals also feel threatened culturally by the Nigerians, many of whom wear traditional African *agbadas* that the São Toméans mock as "pajamas," or practice Islam. And the Nigerians "don't do themselves any favors," in the words of an NGO director I spoke to. They think the São Toméans are lazy and "primitive."

With the anticipation of oil wealth, the São Toméan fear of Nigerians has increased exponentially. Many São Toméans worry that the English-speaking Nigerians will be better able to snap up any jobs that the international oil industry creates on São Tomé, and that they will settle in the country and eventually overwhelm the native population. These São Toméans see the lopsided JDZ agreement and the petty market traders as a pincer movement of Nigerian dominance, performed with the full complicity of São Toméan politicians.

Their fears are largely justified. Nigerians are among the most experienced oil negotiators in Africa, having created a national oil company that operates 55/45 joint ventures with foreign multinationals rather than the far-more-exploitative Production Sharing Agreements typically entered into by the continent's weaker states, like Equatorial Guinea. And from the beginning, the story of São Tomé's attempts to access its petroleum endowment has been one of Nigerian domination—sometimes crude and sometimes highly sophisticated, but always formidable and seemingly insurmountable.

One of the best illustrations of how Nigerian muscle works in practice has been the allocation of exploration blocks in the JDZ—a convoluted process fraught with delays and behind-the-scenes wrangles that have always left the São Toméans holding the short end of the stick. During the first round of bidding in 2004, for example, a consortium led by Chevron won the rights to explore in Block 1 of the

JDZ, but not before an obscure Norwegian independent called Energy Equity Resources was given a 9 percent equity stake in the operation. It turned out that EER was a vehicle for Eliko Dangote, a Nigerian sugar, cement, and groundnut magnate with close ties to President Obasanjo.

Nigeria's strategy in São Tomé is to make it possible for Nigerian companies to link up with more-experienced foreign players, from whom they can learn the business of deepwater offshore drilling in order to compete on a level playing field when future blocks come up for tender. For years Nigeria has required foreign majors to partner with local companies as part of a drive to include so-called "local content" in their operations, but this has happened only in the decrepit downstream sector of the Nigerian oil industry. "If Nigerian companies can position themselves in the JDZ," says Philippe Vasset, editor of the Paris-based newsletter *Africa Energy Intelligence*, "then U.S. companies will be obliged to work with them." São Tomé, he says, is "virgin territory for Nigerian political influence. It's where the dream of local content can be realized for the first time."

The Block 1 bidding round had concluded in late 2004, and São Tomé had believed its 40 percent stake in the signing bonus ($49 million) would arrive in early 2005 and could be factored into the state's 2005 budget. As the second round of bidding on JDZ oil blocks got under way, São Tomé and Nigeria ran into sharp disagreements over which companies would get the remaining blocks, with the Nigerians wanting to give preference to a number of small and relatively inexperienced Nigerian independents. São Tomé balked, and Nigeria began playing hardball, finding ways to delay payment of Chevron's $123 million signature bonus to its JDZ contract. As 2005 wore on, the São Toméans realized they would have to capitulate. The Nigerian government could afford to wait for a few million dollars, but to São Tomé, with its $50 million annual budget, a $49 million bonus was a matter of life and death. Fired up by the prospect of an inflated national budget, state employees in São Tomé had already gone on a week-long strike demanding their monthly salaries be increased from $30 to $100. Schools, ministries, and hospitals were shut down.

On May 31, 2005, therefore, after a five-month delay, the second-round winners were announced. Not surprisingly, such obscure Nigerian companies as Momo Oil and Godsonic Oil & Gas (both considered fronts for controversial Abuja-based businessmen close to the president) took up substantial equity stakes alongside international operators such as Anadarko, Devon, and Noble. Fifteen percent of Block 2 went to Equator Exploration, run by the flamboyant Ukrainian-Canadian businessman Wade Cherwayko, a friend of the Trovoadas. Amid all the monkey business, ExxonMobil declined to exercise its preferential rights and, later that year, Devon and Noble also pulled out, casting serious doubts over the future of the JDZ.

At the second-round signing ceremony in Abuja, President de Menezes went out of his way to paper over the cracks in São Tomé's relationship with Nigeria, saying that "our worries and questions were only and shall always be questions and clarifications. We just wanted to know as a curious younger brother why our big brother is doing this because we are younger in this industry. Nothing whatsoever apart from curiosity and anxiety." It was a splendid bit of groveling, but de Menezes had already confided to a foreign NGO official that dealing with the Nigerians and the international oil companies over the JDZ sometimes felt like "being a spectator at a tennis match," and it is easy to see why.

To a great extent, Nigerian interests have been able to gain such a foothold in São Tomé because of the decline of Angola as a player in the country's political scene. Angola and Nigeria are among only a handful of African countries that can really project beyond their own borders militarily. From 1975 to 1991, the ruling Movement for the Liberation of São Tomé and Príncipe (MLSTP) was closely allied with the MPLA in Angola. Both parties were composed of Marxist revolutionaries who had cut their teeth in the struggle against the Portuguese, and both had come to power with the fall of the Salazar regime in Lisbon in 1975. The personal links between the leaderships in Luanda and São Tomé were also very strong: São Tomé's first president, Pinto da Costa, used to chase girls with José Eduardo dos Santos when they were young men, and the two remain close friends.

Until the early 1990s, Angola even stationed troops on the island as a form of protection for its weaker cousin. In 1991, however, both countries converted to multiparty democracy and Angolan influence began to wane in São Tomé. In part, this was because São Toméan voters threw the MLSTP out of office in favor of Miguel Trovoada—who not only had Nigerian connections but also allegedly had links with UNITA rebels in Angola. But it also had to do with the MPLA's abandonment of doctrinaire Marxism and the coming to power in Luanda of a younger generation increasingly less interested in the rhetoric of international solidarity.

Recently, however, as the Gulf of Guinea has become a potential petroleum bonanza, and as Angola emerges from its brutal civil war, Angola has been challenging Nigeria for political and military dominance in the Gulf of Guinea, and especially São Tomé. A massive new Angolan embassy is being built on the beach in São Tomé town, and the Angolan NOC Sonangol has offered its technical and negotiating expertise to the São Toméan oil industry. In late 2006, São Tomé even invited Angola's infamous "Ninja" police units to come and train its own police force. A genuine rift has opened up among São Tomé's politicians over whether the country should allow itself to drift from an Angolan sphere of influence into a far-less-familiar Nigerian one, and a bitterness over São Tomé's role as a Ping-Pong ball between the two powers was obvious when I met the country's former prime minister, Joaquim Rafael Branco, at his office in the National Petroleum Agency in São Tomé. Branco, an MLSTP man, was blunt about what he saw as de Menezes' failings. "Nigeria has growing influence here," he said. "You have to counterbalance that with Angola. He has destroyed relations with Angola."

It is true that de Menezes has never developed a close working relationship with dos Santos, but he has also worked hard to project an image of himself as an independent politician, unafraid to pick fights with the Nigerians. For much of his first term, his strategy appears to have been to court powerful third parties, in particular the United States, in the hopes of getting both regional powerhouses to back off.

This strategy has suited the Americans very nicely. The U.S. defense establishment has wasted little time in touting São Tomé—which it sees as a weak and potentially loyal ally with an extremely attractive geographical situation in the middle of the Gulf of Guinea—as a country worth doing business with. In September 2003 the deputy commander of the U.S. military's European Command (EUCOM), General Charles Wald, suggested that a nonpermanent American base could be built on São Tomé as a way to ensure stability in the Gulf of Guinea. In March 2004, the private military contractor MPRI sent retired General Raoul Henri Alcala to São Tomé as a consultant to conduct a one-year coordination of security-cooperation projects, including military education and training, along with equipment sales and transfers. The effort focused on the São Toméan coast guard, with the idea that São Tomé might eventually help patrol the waters of the JDZ. In May 2004 the U.S. military followed up on MPRI's efforts with a weeklong workshop on civil-military relations, and in August President de Menezes and his defense minister visited EUCOM's headquarters in Stuttgart. Later that month, General Wald returned the visit, bringing with him Senator Chuck Hagel, who sits on the powerful Senate Foreign Relations Committee. (The two also visited Nigeria, Angola, Gabon, and Cameroon.) Wald sat across from de Menezes and listened to a long list of São Tomé's urgent infrastructure and civil-defense needs—highest among them a lengthening of the landing strip at the country's decrepit airport and the construction of a deepwater port. By the end of 2004 American aid to São Tomé had been doubled (although it remained quite small at $296,000), and it was announced that the United States would fund a feasibility study for the building of a new port.

In the American press, São Tomé was hyped as the linchpin of an American strategy to diversify U.S. oil supply while much of the international (and left-leaning American) media panicked over what they saw as more gunboat diplomacy coming out of Washington. EUCOM had indeed taken unusual initiative for a Combatant Command center and arguably acted beyond the scope of its mandate. Frustrated Pentagon officials found themselves repeatedly having to deny that the United States had plans to build a base in São Tomé, with some privately describing the activist posture of EUCOM as the product of generals with too much time on their hands.

In truth, by the late 1990s, EUCOM, which had been set up after World War II to maintain the U.S. security arrangement in Western Europe, had begun to find itself casting about for ways to maintain its relevance. EUCOM was beginning to show its age. With the relatively isolated exception of the Balkans, Europe seemed unlikely to explode into a theater of warfare any time soon, and both NATO and the European Union were taking a more proactive role in European security than they had in the past. The U.S. military's awkward definition of "Europe" had been extended in 1983 to include most of sub-Saharan Africa, but EUCOM's commanding officers had never taken an enormous interest in the continent found just below Europe on their maps. Not, that is, until ACRI came along.

The Africa Crisis Response Initiative, or ACRI, was the brainchild of the State Department, but as it was a military and logistics initiative, its implementation fell to the Pentagon, in the form of EUCOM. Officially, ACRI was meant to train African militaries in peacekeeping and humanitarian work. Any military hardware provided by the United States would be of the nonlethal variety, such as generators, vehicles, and night goggles. Many, though, saw ACRI as the Clinton administration's way of preparing African militaries to deal with the threat of terrorism, as well as ensuring against the possibility of the United States getting sucked into the security vacuum left by another failed state, as had happened in Somalia in 1993. In charge of the initiative was Colonel Nestor Pino-Marina, a Cuban exile who had participated in the failed 1961 invasion of the Bay of Pigs, and gone on to be a Special Forces officer in Vietnam and Cambodia, before leading covert operations against the Sandinistas in Nicaragua. As such initiatives go, ACRI was small, but it gave EUCOM's brass a taste for the seemingly limitless complex security challenges that characterize the African continent.

Then came September 11, 2001. The U.S. government suddenly realized no area of the world could be taken for granted anymore. In response to the terrorist attacks on American soil, the Bush administration launched Operation Enduring Freedom as part of its Global War on Terror. In spring 2002 the administration announced that ACRI would be reorganized and rejuvenated, and given the name ACOTA—Africa Contingency Operations Training Assistance. Gone was the fig leaf of humanitarian work, replaced by a robust emphasis on "contingency planning." Unlike ACRI, ACOTA would include training for offensive military operations, including light-infantry tactics and small-unit tactics, to enhance the ability of African troops to conduct peacekeeping operations in hostile environments. Under ACOTA, African troops would also be provided with offensive military weaponry, including rifles, machine guns, and mortars.

By 2003 EUCOM's commanders had grown increasingly convinced that Africa was an important front for the United States in the new global war. In November that year, they put together a little program of their own, called the Pan-Sahel Initiative (PSI). Very small by military standards, the PSI represented a departure from traditional cooperation programs in that it focused on a particular region—the vast, sparsely populated band of semiarid and frequently drought- and famine-stricken land known as the Sahel that marks the transition between North Africa and sub-Saharan Africa. In 2004 the PSI worked with four Sahelian countries—Chad, Niger, Mali, and

Mauritania—to combat smuggling and take action to prevent cross-border crime and terrorism, at a cost of $6 million.

In 2005 the PSI was rebranded the Trans-Sahara Counter-Terrorism Initiative, or TSCTI. Its annual budget was increased to $100 million, and another six nations became involved. Thus, EUCOM, acting largely on its own, extended the Global War on Terror to the Sahel. Many believe that, given the opportunity, they would go farther, faster.

While EUCOM's activities in the Sahel were sanctioned by the Departments of State and Defense, EUCOM's commanders began to feel there was another, equally important aspect of African security being neglected by official Washington. As early as 2002, Stuttgart had taken note of the Gulf of Guinea's potential importance to American energy security, and was talking in animated terms about the fact that the oil-rich 2,000-mile coastline along the Gulf, dotted with ever-more-prolific drilling platforms and offloading vessels, was mostly unpoliced. In October that year, EUCOM's Deputy Commander Carlton Fulford visited São Tomé, amid much buzz about the possibility of a U.S. base. Fulford's visit, which appears to have been primarily on his own initiative, caused some embarrassment for the Pentagon, which was by now denying plans to build any such base.

But EUCOM charged ahead.

In 2003 EUCOM proposed the creation of a Gulf of Guinea Guard, modeled on past U.S. Coast Guard assistance programs, and in October 2004 it invited chiefs of naval operations from the Gulf of Guinea nations to a Coastal Security Conference in Naples. Remarkably, it was the first time they had met as a group, and out of that conference came a commitment to improving security cooperation in the region. EUCOM went further, proposing a regional maritime control and surveillance center, and arguing that in a larger effort to address the poverty and lack of accountability that were the root causes of conflict and lawlessness, international oil companies would need to step up their socially responsible investments, and France and Britain would have to be engaged. In November 2006 a follow-up security conference was held in Benin, with eleven Gulf of Guinea countries in attendance, and, in January 2007, the U.S. navy installed $18 million of surveillance equipment in São Tomé, the first phase of its Regional Maritime Awareness Center for the Gulf of Guinea.

EUCOM's freewheeling approach and liberal interpretation of its mandate complicated efforts by the U.S. military to heighten its profile in Africa. The Command's activities have sometimes trodden on delicate sensitivities in the host countries, or been viewed with suspicion. In March 2005, in the Nigerian capital, Abuja, the Pentagon sponsored a weeklong seminar on energy security and the Gulf of Guinea. There, American commanders and their African counterparts had what in diplomatic parlance is referred to as "a series of frank exchanges." At issue was the deteriorating security situation in the Niger Delta, where Dokubo Asari's activities had drawn attention to the militias stealing millions of dollars' worth of crude oil and using the money to buy weapons for an increasingly ugly insurgency against the Nigerian government.

The African commanders suggested that countries that depend for their survival on imported energy (i.e., the United States) should bear some responsibility for funding a rapid expansion of local navies to help protect that oil. The Americans strongly disagreed, stating that corruption and official complicity with crude-oil smuggling were problems that had to be addressed first. Some of the Pentagon's sharpest disagreements were with Admiral Samuel Afolayan, chief of staff to the Nigerian navy, who insisted he was helpless to stop the criminal activity without outside help. Nigerian authorities had recently arrested two high-ranking navy admirals in the case of the missing ship *MT African Pride,* which EUCOM's Carlton Fulford acknowledged as an encouraging sign. "But," he added bluntly, "it probably goes higher than that." This swipe at the ethics of Nigeria's leaders inflamed the hosts of the gathering. But Fulford stood his ground. Illegal crude bunkering was "not an international security problem," he insisted. "It's a Nigerian issue, and they have to deal with it. Nigeria is my single biggest concern."

Those inclined to view U.S. military activity as inherently suspicious have drawn from EUCOM's activities an overarching narrative of American imperialism and big sticks, but this is giving in to a cynical and conspiratorial view of the world. For starters, EUCOM's excitement about Africa—and its tendency to act without consulting the Pentagon—hasn't always received the warmest reception in Washington. According to diplomatic sources, several visits EUCOM commanders made to Africa were informal and unauthorized, because the generals knew that authorization from Washington might never arrive. This has sparked an extended interagency debate in Washington about exactly how vulnerable offshore oil installations in Africa really are, and how high up the Pentagon's priority list they should be. The Departments of State and Defense, in particular, have felt unable to justify diverting resources from more pressing problems, such as the Iraq war. Nevertheless, in February 2007, President Bush gave the green light to a Pentagon plan to create a dedicated Combatant Command for Africa—a sort of "AFRICOM."

Still, there are real reasons to reject the conclusion that the U.S. military is seeking a heavy footprint in Africa. Even at $100 million, the Trans-Sahara Counter-Terrorism Initiative is underresourced, given the ambitiousness of its intentions. The landmass represented by the ten countries the TSCTI wants to help police is larger than the continental United States, and almost entirely desert. Much the same point can be made of the much-touted Gulf of Guinea Commission, which is supposed to help the region's navies and coast guards coordinate their activities. As for the rumor that the United States was about to declare the Gulf of Guinea an area of top priority and start building a base in São Tomé, it almost certainly owed its existence to the publicity efforts of Paul Michael Wihbey. Ultimately, a lot of the American security interest in the Gulf of Guinea comes down to what one analyst calls "heavyweight policing," especially of the criminal networks centered in Nigeria.

If there is any cause for concern, it is not EUCOM and a few opinionated generals, but who the U.S. Defense Department has found to do the heavy lifting in the Gulf of Guinea. Today much of America's military footprint overseas is not directly supervised by the Pentagon, but by private military contractors. Military Professional Resources International—a Virginia-based "professional services company" run by retired generals and diplomats that claims a core commitment to defending "the values that are at the very foundation of our nation"—has provided staff and technical support to the Coalition Provisional Authority in Iraq since the U.S. invasion in March 2003. Since 2002 MPRI has also been developing and implementing an "action plan" for Afghanistan that includes the formation of a modern defense ministry and the organization and training of that country's new army and air force. Much of the training of African troops that took place as part of ACRI and later ACOTA was conducted by MPRI.

But MPRI's activities have gone far beyond implementing ACRI and ACOTA. The company has taken a particularly high-profile role in Nigeria, where it has found itself at the center of heightened tension between the United States and Nigeria. When Olusegun Obasanjo was elected the first civilian president of Nigeria in 1999, he was eager to demonstrate to the world that the military would no longer play a role in the nation's politics. Obasanjo signed an agreement allowing the United States to implement a training program designed to "reprofessionalize" the Nigerian military, making it possible for them to "disengage" from civil-government functions. (Cynics in both the United States and Nigeria guessed this "transition assistance" was inspired by an American desire to ensure the stability of its fifth-largest oil supplier.) The contract for the work was awarded to MPRI, which quickly ran into problems with the Nigerian top brass.

The Nigerians felt they had performed with distinction in peacekeeping operations in Liberia and Sierra Leone and had nothing to learn from MPRI, which at least one general accused of being a nest of spies. When MPRI's contract expired in 2003, the Nigerian government refused to renew it, despite heavy American pressure. In retaliation, the United States renewed a ban—lifted at the end of the military dictatorship—on Nigerian military personnel receiving training at American institutions.

MPRI's most controversial African intervention, however, has been in Equatorial Guinea. In 1999 MPRI applied for a license from the State Department that would allow the company to provide services to the Obiang government. The request was refused repeatedly on the grounds of Equatorial Guinea's poor human-rights record. MPRI lobbied Congress, asking it to put pressure on the State Department to reconsider its decision, and a license giving MPRI permission to train the Equatoguinean coast guard was eventually granted. Despite vocal opposition from some lawmakers concerned with U.S. military training being used by a government that terrorizes its own citizens, MPRI was also given the go-ahead to begin a more comprehensive National Security Enhancement Plan for Equatorial Guinea in early 2006.

◄◄·►►

Even if São Tomé has been oversold and overhyped, big changes are clearly on the way for its people. After all, only a few thousand barrels of oil a day would bring about a complete transformation in this minute archipelago with a population the size of Providence, Rhode Island, and a GDP of just $69 million. Little surprise, then, that the Portuguese group Pestana has begun building a five-star hotel and casino complex in São Tomé that will include a discotheque, holiday villas, and long-term housing for expatriate businesspeople, to be completed in 2008 at a total cost of $17 million. Other Portuguese construction firms have put up offices and apartments in anticipation of the arrival of Chevron staff. As of 2005, the country was receiving barely five thousand tourists a year, most of them Portuguese, who have long enjoyed one of the world's best-kept secrets. But the government has identified tourism as a priority sector and has big plans for attracting the European holiday market. Forecasts suggest that private investment in the hotel sector alone could reach $27 million in 2008, a staggering sum by São Toméan standards.

The islands' rampant malaria and lack of a decent hospital may hurt São Tomé's future as a tourist destination. Probably the biggest bottleneck, though, would be the lack of flight connections. Most visitors currently arrive on an expensive weekly Air Portugal flight from Lisbon. Until recently, the only other option was a short but nerve-rattling hop from Libreville onboard a threadbare De Havilland Twin Otter operated by the virtually bankrupt Air São Tomé and Príncipe (an experience I won't soon forget). In May 2006, though, this option was taken away when Air STP's only plane crashed into the ocean, taking the lives of eleven passengers and crew. This latter hurdle is one the government feels it can readily overcome, however, and it seems the days of deserted beaches and cozy eccentric hideaways on the misty slopes of volcanoes may soon be over.

As São Tomé prepares to open up to the world of petroleum markets and deepwater drilling, one of the biggest challenges for the government will be handling the oil wealth responsibly while managing the expectations of the population, many of whom believe they will soon be driving BMWs. To this end, São Tomé has brought in the renowned international-development guru Jeffrey Sachs, who, along with a team of experts and students from Columbia University, has been putting together a "Development Plan of Action" for the management of São Tomé's future oil wealth. The Columbia team, working pro bono, helped São Tomé write an Oil Revenue Management Law that its parliament passed unanimously in November 2004. Among its highlights are the establishment of a permanent fund dedicated to development projects and poverty reduction, and a

Norwegian-style fund for future generations, meant to ensure that São Tomé does not suffer unduly when the oil runs out. São Tomé is supposed to become a "model" emirate—an African oil producer that turns its natural resources into a blessing, rather than a curse.

As promising as this all sounds, however, it is hard to escape the nagging feeling you get about a country so lacking in institutional capacity that it entrusts the management of the most important social and economic transition it has ever gone through to graduate students working for free in their spare time. Surely what Africa's newest oil producer needs most is some heavyweight technical assistance and a carefully arranged transition program of institutional capacity-building?

On one of my last days in São Tomé, I rented a 4x4 and drove to some of the island's more remote villages to see what the population outside the capital felt about the coming of oil wealth. I stopped in Agua Izé, a handful of tidy cobblestone streets flanked by faded stucco houses sitting on a plain overlooking the palm-fringed coastline. The village was once part of a plantation estate, and at the top of the hill stood the old plantation hospital, a magnificent stone edifice with a grand twisting balustraded staircase in front that was being steadily swallowed up by tropical vines and banana trees. The year of its construction—1914—had been chiseled elegantly over the entrance, a reminder of a time when the Portuguese empire believed it was here to stay. Now, a woman and her children, dressed in filthy T-shirts and sandals, sat in the empty window frames looking bored.

In the cobblestone streets below, a toothless old man played a traditional Portuguese fado on his guitar while a group of young men sat on piled-up planks of wood arranged around a fire, watching a pot of water boil. My presence attracted a crowd of angry young men who were more than willing to share their thoughts about the promise of oil in São Tomé. Despite their tattered clothes and beer-infused breath, they were articulate and well-informed about the *"dossier petróleo,"* as the oil issue is called in São Tomé. All were aware that it would be several years before the country actually saw any money from oil exploration, and all expected their elected politicians to squabble over access to the spoils. Few expected any real change to their lives.

"Look around you!" one man shouted, and I paused to take in the grass growing through cracks in the stucco walls and from between the cobblestones. Children and pigs and chickens wandered about, ducking in and out of houses whose doors and windows had been missing for decades and whose roofs had been badly compromised by years of tropical rains. "The schools and hospitals we are using are the same ones the white man left behind. We keep hearing money is coming, money is coming, but it is just coming and going to banks in Switzerland and the United States. Oil can be a blessing or a curse, and if São Tomé turns out to have no oil, we could all end up owing too much money."

A little farther down, at the end of the island's only paved road, is the town of São João dos Angolares. There, a Portuguese family had converted the small colonial plantation house into a tasteful bed-and-breakfast and art gallery. Fresh coffee beans from the estate's trees were being ground by an old man using an antique hand-cranked mill, filling the veranda restaurant with an intoxicating aroma. Local artists' paintings and sculptures were on display. São João was, in short, possibly the most romantic place on earth.

Was it possible—just possible—that the African oil boom didn't have to be a story with an unhappy ending? Would this tiny twin-island nation, with its open democracy and stated commitment to revenue transparency and sound fiscal management, prove the cynics wrong? Even good people tend to go a bit funny in the face of millions of dollars, and there was no compelling evidence that São Tomé's politicians were "good people." In an impoverished and desperate country with weak institutions, powerful neighbors, and a culture of dinner-table politics, there is every likelihood that a spectacular infusion of foreign exchange will act as a destructive and destabilizing force. But, say the optimists, at least São Tomé has the advantage of coming late to the game, and having the examples of Gabon and Nigeria and even the relative newcomer Equatorial Guinea to learn from.

Perhaps the optimists will be proved right, and São Tomé will become a "model" for every African country hoping to benefit from the continent's oil boom. Unfortunately, those who follow African oil have heard it all before—not in the leafy, palm-fringed volcanic islands of the Gulf of Guinea, but more than one thousand miles from São Tomé, on the parched and cracked canvas of dust and sand where the jungle meets the Sahara. There, a few short years ago, the Republic of Chad—a basket case of war and hunger and tribal hatreds—was being hailed as a potential "model" for the responsible management of oil resources in an African country. Despite being one of the poorest, most corrupt, and most brutal places on earth, Chad was going to build a pipeline and show Africa that oil could be a blessing rather than a curse for a nation and its people.

Did it work? That depends very much on whom you ask. But, as I was about to find out, the signs were not good.

CHAPTER 6

THE PLACE WHERE PEOPLE WAIT

TRYING TO TALK in general terms about the vast continent that has been referred to for millennia as "Africa" is a notoriously difficult task. For the sake of simplicity, historians, geographers, political scientists, and Africanists of many other disciplines have tended to talk in terms of Northern and Southern Africa, the former characterized by the harsh landscape of the Sahara desert and by the Arab civilizations that have dominated it since the seventh century, and the latter by a mixed landscape of tropics, savannas, and highland plains populated mostly by Bantu-speaking black "Africans." Like all generalizations, though, this leaves out what doesn't fit. Most of us think we have a pretty good idea of what North Africa looks like—mosques and minarets, Bedouins and date palms, pyramids and pharaohs—and a pretty good idea of what sub-Saharan Africa looks like—leopards and lemurs, plantains and baobabs, Afrikaners and ivory—but are hard-pressed to bring up a mental picture of what happens where those two worlds meet.

When the ancient Greeks spoke of "Africa," they meant the North Africa of Berbers and Copts and Libyans. What lay on the other side of the Sahara they referred to as *Æthiopia*, or "land of the Blacks." When the Arabs took North Africa in the Middle Ages, they referred to the unconquered terrain to the south as *Bilad as-Sudan*, or (you guessed it) "land of the Blacks." And as for the native Berbers farther west (around present-day Morocco and Algeria), they referred to the land past the Sahara as *Aral n-Iguinawen* (which the Portuguese exploring the west coast of Africa later bastardized into "Guinea"). No prizes for guessing what it meant.

We know less about what sub-Saharan Africans thought of their lighter-skinned neighbors to the north, but we do know that for centuries, the thin band of dry, dusty, grassland that separates the Sahara desert from "black" Africa has been the scene of fighting, slavery, droughts, famines, plagues, and one of the most precarious existences known to the human family. When the rainy season comes to the Sahel—as geographers now refer to the area—it often brings no rain, or the rains come too late for the growing season. Other times it rains so hard that all the seeds wash away. Even when the weather is cooperating, swarms of locusts can wipe out crops overnight. In the last fifty years alone, millions of people have died as a result of repeated crop failures in the Sahel. The intense daily struggle to survive in these harsh conditions has triggered centuries of competition and tribal battles over water and grazing rights. And with the introduction of modern states and well-funded armies in the 1960s, the Sahel has turned into a living hell—a land of instability and famine and genocidal warfare, where millions of people survive by gnawing on twigs and desert rats and waiting for the next sack of grain delivered by the white people in airplanes. It is truly one of the most miserable places on earth.

And right in the middle of it stands the Republic of Chad—a crudely drawn swatch of sovereignty tugged at from north and south and east and west and, probably most of all, from within. Hundreds of miles in any direction from the great oceans of the world, Chad is a hot, choking, neglected tragedy of a nation that ranks consistently at the bottom of virtually every indicator of human development. In a country twice the size of France, there is less than 400 miles of paved road, one domestic flight route, and no railway. There is less than one telephone for every twenty people, less than one car for every one hundred people. More than half of Chad's adults are illiterate, and three-quarters of the population lives on less than one U.S. dollar a day.

For most of its history as an independent nation, Chad has been torn apart by ethnic, religious, and political violence. The country's boundaries take in a wide swath of the Sahara Desert in the North—home to nomadic dark-skinned (or "Africanized") Arabs reliant on cattle herding—and stretch into the semiarid and occasionally tropical South—home to dozens of "black African" tribes—with a thin middle belt of Sahelian grassland in between. During the period of French rule, from 1920 to 1960, black southerners were encouraged to dominate politics in the capital, N'Djaména, a dynamic that continued into the early years of independence. But the radically different cultures of North and South soon began to squabble over the country's paltry resources. For much of the 1960s and 1970s, the country was devastated by one of Africa's worst civil wars.

In January 1981, just as a tenuous government of national unity had begun to deliver a North-South power-sharing arrangement and a measure of peace to the country, Chad's new president, Goukouni Oueddei, made a costly tactical error, announcing that Chad and its northern neighbor Libya intended to "merge" gradually into a single nation. The newly elected Reagan administration had made the containment of Libya's Muammar Qaddafi a top priority, and began giving massive covert support to a Chadian northern splinter group, the FAN, led by Hissène Habré. In June 1982 Habré seized power in N'Djaména and, supported heavily by both France and the United States, terrorized Chad's population for eight long years. During Habré's presidency, some 40,000 political opponents were assassinated by the secret police—the dreaded DDS—while Libyan-backed forces continued to occupy the country's extreme north.

The DDS specialized in creative techniques for extracting confessions and information. Electric shocks, beatings, whippings, and extraction of fingernails were all routine, but victims could also expect hot pepper gas to be blown through pipes pressed against their temples, or to have lit twigs and matches held against their most sensitive body parts. Sometimes large quantities of water would be forced down the throats of victims before DDS agents stomped onto their stomachs. One of the most extreme forms of torture included forcing the exhaust pipe of a running vehicle into the mouth of a victim. Simply accelerating the motor would cause severe burns.

In 2005 Human Rights Watch uncovered documents revealing the depth of U.S. support for Habré, who in 1987 was received at the White House and toasted by policymakers as a "friend" of the United States. President Reagan welcomed Habré's resistance against the "violent aggression of an outlaw state [Libya]," and praised his "commitment to freedom and international cooperation." In 1985 some of Habré's most senior DDS agents had been flown to a location outside Washington, where they received "very special" training from "our American friends," whom they described as "attach[ing] a very high degree of importance to this training." U.S. agents even promised to supply "equipment" to their Chadian counterparts. Documents refer to Chadian requests for "truth serum" and a "generator for interrogations."

It would take an eccentric sensibility to find anything beautiful or sublime about Chad. It has inspired no Wordsworths, no Shelleys, no brooding poets wandering the Sahel for inspiration. The more common human response to the landscape has been to try to shut it out of sight and out of mind, to pretend it doesn't exist and hope it goes away, and to focus instead on surviving the sadistic extremes of heat and dust it produces. In the south of the country, villagers build small circular huts out of dried mud and straw, with no windows and only a small opening for a door. In the north, where sandstorms are a daily nuisance, men wrap their faces and heads in long white scarves and cover their eyes with aviator sunglasses, so their features become completely invisible. Even the French were unmoved by this brutal landscape, referring to the area south of the Chari River, where the semitropical soil was well suited to cotton farming and other forms of agriculture, only as *Tchad util* (Useful Chad). The other 90 percent of the country, presumably, was Useless Chad.

In 1996 *Tchad util* turned out to be more useful than anyone had imagined. A seismic-exploration program run by ExxonMobil confirmed the presence of between 800 million and 1 billion barrels of crude oil in the Doba basin and nearby formations in southern Chad. Foreign companies had been interested in Chad's potential as an oil producer since the late 1960s, and in the late 1970s the American major Conoco had even drilled exploratory wells, but nothing much had come of it. Chadian crude is of the heavy and sour variety that fetches low prices on the international market, and the country's land-locked geography adds formidable transportation costs to any venture. Besides, with civil war and political instability a fact of life from 1965 until the early 1990s, there was never much chance of Chad's oil industry getting off the ground. In 1996, however, there seemed to be just enough oil in Chad, and nearly enough political stability, to justify giving the country another look. ExxonMobil began to examine financing and feasibility options, setting into motion what would become one of the most extraordinary chapters in the history of African oil exploration.

Exxon quickly realized that the obstacles would be enormous. Chad's appalling infrastructure meant that everything would have to be flown in and built from scratch. Exxon would have to erect a company town—complete with its own power plant, airport, and water-treatment facilities—in the middle of the Sahel. Then it would have to lay down a 660-mile pipeline from southern Chad through the length of neighboring Cameroon, all the way to the port of Kribi on the Gulf of Guinea. The pipeline would run through virgin Cameroonian rain forest, inhabited by reclusive tribes of Pygmies, and almost certainly disrupt delicate ecosystems. The price tag for this little adventure was likely to run into the hundreds of millions of dollars, if not stretch to several billion, and to come with a PR headache.

In 1996, in fact, Africa was the last place an international super-major wanted to be seen doing business. It had only been a few months since a Nigerian military junta had hanged Ken Saro-Wiwa and eight other environmental activists from the Ogoni tribe for a murder few believed they had committed. Rightly or wrongly, a wide spectrum of European civil-society groups believed that Shell, the largest operator in Nigeria, had had direct complicity in the activists' kangaroo trial, and the eighty-eight-year-old multinational had been forced to beat a humiliating retreat from Ogoniland. Moreover, memories of the Rwandan genocide, in which 800,000 Tutsis and moderate Hutus were killed in weeks, were still fresh. This catastrophic human tragedy had nothing to do with Big Oil, but it hardly did wonders for Africa in the Western imagination. To think about going into Africa in the mid-1990s, an oil company had first to recognize that the faintest whiff of complicity with political violence or an inability to address the grievances of the natives was a public relations disaster waiting to happen.

Not wanting to sink billions into a risky venture whose commercial value was untested, ExxonMobil took the unusual step of approaching the World Bank to discuss the possibility of receiving funding for the project, as well as the kind of political legitimacy only an international institution like the Bank could provide. Created at Bretton Woods in 1944 with the aim of alleviating global poverty, the Bank didn't see supplying financing for a large multinational corporation like ExxonMobil as part of its job. But in the mid-1990s, it was casting about for a showcase project to demonstrate that its philosophy of loan conditionality and economic liberalization was still the best way to improve lives in the Third World. The Chadian government was in no position to resist. Emerging from decades of civil war and instability and totally dependent on international aid, Chad was, in the parlance of the international financial institutions, a country where the Bank could exercise a considerable amount of "leverage."

For years, international financial institutions such as the Bank and the IMF had, controversially, imposed conditions on heavily indebted countries and asked them to demonstrate progress toward sound fiscal management practices before helping them with debt relief or development loans. They had offered armies of technical advisers and consultants to "help" with these transitions. But never before had they had the chance to lay out chapter and verse how a government could and could not spend its money.

As a condition of its support for Exxon's project, the World Bank required Chad to impose strict measures on the management of oil revenue. Royalties from the sale of oil would be paid directly into a Citibank escrow account in London, rather than into the Chadian treasury, where it risked disappearing. (In 2005 Transparency International ranked Chad as the world's most corrupt country.) Citibank would invest 10 percent of the money in a "Future Generations Fund" so that Chad wouldn't turn into another Gabon when the oil ran out. The remaining 90 percent would go back to Chad through closely monitored "Special Petroleum Revenue Accounts" and be divided according to a formula that reserved 80 percent for "priority sectors"— education, health, rural development, infrastructure, and environment—and 5 percent for the oil-producing Doba region itself. Only the remaining 15 percent (or 13.5 percent of the total) would be available for the government to spend as it saw fit (most likely on salaries and weapons—the building blocks of a weak regime's survival). Moreover, the process of allocating revenue to priority sectors and local development would be overseen by a Petroleum Revenue Oversight and Control Committee (known as the *Collège*), composed of representatives from civil society, trade unions, the government, and the churches. On December 30, 1998, after only three hours of debate, Chad's National Assembly passed the proposed measures, which it promulgated as Law 001 of the Republic of Chad.

It is hard to think of another moment in history when a sovereign country has allowed foreign players to dictate the management of its internal affairs to such a level of detail, during peacetime, with barely a peep of dissent. It should be borne in mind that, as a desperately poor country, Chad had already agreed to let Exxon keep 100 percent of the profits from the Doba project and collect only royalties, so in effect,

Law 001 was dealing with the 12.5 percent royalty payment that the Chadian government collected. In other words, the 13.5 percent of "revenue" that the government was allowed to use at its discretion was actually less than 2 percent of the total profit from its own oil fields. Between ExxonMobil's savvy negotiating skills and the World Bank's strict conditions, Chad's leaders were treated as little more than spectators to the country's oil boom, and were furious at the humiliation. However, given the country's history of high-level corruption and repression, few honest observers really objected at the time. From the World Bank's perspective, Law 001 was a dream come true—a rare opportunity to test the institution's economic wisdom under almost-laboratorylike conditions. For the Bank, a lot was riding on the Chad experiment. If it failed, its armies of omniscient experts might, for once, find themselves with no one else to blame.

Almost immediately, the experiment ran into problems. Chad's president, Idriss Déby, spent part of an initial signature bonus of $25 million from ExxonMobil on military hardware to fight rebels. The World Bank was forced to acknowledge that, technically, Law 001 covered only profit oil and not signature bonuses. Déby was on the legal—if not the moral—high ground, but the Bank's then-president, James Wolfensohn, was having none of it. "What the hell do you think you're doing?" he is reported to have shouted when Déby answered his phone in N'Djaména.

Civil society, too, was skeptical from the beginning. Local NGOs and religious leaders in Chad (egged on by German and American advocacy groups) began to educate themselves about the experiences of activists in Nigeria and elsewhere in Africa, even paying a visit to the Ogoni. On the day the valves opened on the pipeline, a group of Chadian NGOs proclaimed a "national day of mourning." It didn't help matters that ExxonMobil's junior partners in the consortium— Shell and TotalFinaElf—brought with them considerable political baggage. Shell's struggles with the Ogoni had become notorious, and the details of Total's "Africa system" and its involvement in perpetuating conflict and indebtedness in Angola and the Congo were just beginning to emerge. Chadians already had a visceral distrust of French officialdom and felt ill at ease over Total's entry into the country.

The Chad-Cameroon pipeline project was off to a shaky start.

In 1999 civil society in Chad scored what it considered a victory when Shell and Total left the consortium abruptly. The companies cited the low price of oil as a main factor, but word on the street was that they had got cold feet about the controversial venture, a view reinforced by a Chadian government spokesman's public complaint that "the sudden nature of these decisions suggest that they are not dictated by economic or technical considerations." Jubilant demonstrations took place in N'Djaména, complete with the spectacle of the French flag being set afire. ExxonMobil, however, pressed ahead, and by 2002 had found new partners—Chevron and the Malaysian state company Petronas. The new consortium was welcomed by many Chadians, who felt its lower profile in the region and Anglo-Saxon composition made it less of a threat, but there were still concerns about how quickly things were moving forward.*

The government portrayed the NGOs as pessimists and Cassandras who didn't want the Chadian people to become rich from their oil. However, at this stage, the main goal of civil society was not delaying the project simply for the sake of delaying it. Rather, critics were concerned about Chad's extraordinary lack of institutional capacity and weak democratic structures, and wanted to see these improved substantially before construction of the pipeline was allowed to go ahead. The government in N'Djaména, much like that in São Tomé, was inexperienced, and had no real competence to negotiate with multinationals. Even worse, in Chad the government was primarily a small group of people close to the president, who had come to

---

*In much of the Francophone world, the term "Anglo-Saxon" is used rather generally to refer to cultures and systems that have British rather than French/Continental roots. It often tends to be shorthand for a model of business and politics favored by Britain, the United States, Australia, Canada, et al., but can also, as here, extend implicitly to former British colonies, such as Malaysia.

power in a military coup. A minority ethnic cabal surrounded by tribal enemies and with only a tenuous grip on power, it had no meaningful interest in promoting democratic accountability. The World Bank acknowledged that Chad's lack of capacity was "all-encompassing and greater than in most sub-Saharan African countries, reflecting the impact of almost three decades of civil strife," but argued that capacity could be built up quickly and in parallel with the construction of the pipeline. Activists warned that, parachuted onto a barely functioning state, Law 001 would go down in history as a nice idea that never worked in practice.

Most observers now agree that the critics were prescient and the World Bank wrong. ExxonMobil charged ahead with construction, and it quickly became clear that the pipeline would be completed well ahead of schedule—and well before anyone in Chad was ready to manage a petroleum economy. In October 2003 the first barrels of Chadian crude began flowing to Kribi, *fully one year* ahead of the original completion date. Presiding over the inauguration of the pipeline, President Déby declared before assembled heads of state and ExxonMobil's general manager for Chad that "Chadian oil will serve peace in Chad, peace with our neighbors, with the rest of Africa and the rest of the world." But even as Doba crude was being loaded onto tankers in the Gulf of Guinea, the government was bickering with the consortium over how to measure production and sales and calculate revenue—matters which, at press time, still have not been fully resolved.

The brave new world promised by Law 001—a world in which oil revenues are spent on building schools and hospitals—was already in jeopardy. The *Collège*, which was supposed to monitor the movement of revenues from Doba into and out of the various oil accounts, as well as "verify," "authorize," and "oversee" spending in the five priority sectors, had been underresourced, underfunded, and undermined from the beginning. Déby initially named his brother-in-law to sit on the committee, an appointment he was later embarrassed into overturning. The government also created forty-three phantom NGOs and activist groups in an effort to groom a friendly civil-society representative. Worse, the *Collège* could not attract the level of funding and expertise it needed to do its job. Although the *Collège* has four full-time permanent technical staff, its nine appointed members—representatives from civil society, government, trade unions, and the churches—are part-time and must pass judgment on complex matters from road construction to international accounting to the pricing of medical technology.

In April 2005, during my visit to Chad, I dropped in on Father Antoine Berilengar, a Catholic priest who occupied the seat reserved for the representative of Chad's religious communities. Father Berilengar had been on the *Collège* for six months and was already appalled by what he had seen. He complained about the lack of an "expertise office" to help the *Collège* make evaluations, citing an example of benches that had been bought for a school. They were made of recycled wood, but had been passed off by the government as virgin wood. "With benches, we can go and have a look ourselves," said Berilengar, "but what can we do about roads and bridges?"

Father Berilengar was not the only one to notice the inability of the *Collège* to prevent corruption. In 2005 civil-society groups, government officials, and World Bank representatives compiled a shocking list of questionable practices. A construction firm, for example, received $360,000 to build a water tower, but never completed the work. Numerous sanitation projects and health clinics were left partially completed or completely abandoned. Major road projects were awarded to a company run by the president's brother. And several government ministries purchased computer hardware and furniture at prices up to triple the market value. This latter practice seemed ubiquitous. When I met one of the four permanent technical staff of the *Collège* over a beer at a roadside bar on the outskirts of N'Djaména, he confided that the government had bought a book for the university library—*Analyse Micro-économique* by J. Lecaillon—and had billed $600 for it. The book, a fairly standard economics textbook, is available on Amazon.fr for €6.95. Not even Amazon can get away with shipping charges like that.

Cynics claim that this "two-speed problem"—the mismatched timetables of Exxon's project and Chad's development—was the result of a calculation on the part of ExxonMobil to limit its exposure to public criticism by keeping the potentially controversial phase of pipeline construction as short as possible and downplaying the scale of its Doba drilling operation. "They preempted the situation," says Oliver Mokom of Catholic Relief Services in Cameroon. "They knew what public pressure was going to be and they wanted to make the whole thing look smaller than it really was. This is also why they finished it quickly." ExxonMobil tried to avoid protracted negotiations over the route of the pipeline and the way in which affected villagers were compensated by presenting illiterate jungle dwellers with what came to be known as the "Sears catalog." Villagers could choose such items as bicycles or plows as compensation for the destruction of their livelihoods, and eventually, as part of what ExxonMobil described as a period of "social closure," were asked to sign documents saying they would make no future claims against the company. Many were given financial compensation, but no banks were set up, and much of the money was spent by villagers on beer. A study conducted by one local NGO showed that only 9 percent of the money paid as compensation for destroyed crops was reinvested in agricultural development. "They gave cash to Pygmies," says Mokom. "Pygmies live in the jungle on wild fruits and game. Cash payments mean little to them. They need to be resettled, they need schools, they need piped water. It's being done, but very slowly."

But the two-speed problem was only the beginning. International NGOs and Chadian civil-society groups maintain that Law 001 is full of loopholes and inherent weaknesses. For starters, its provisions apply only to direct profit from oil and not to indirect revenues, such as taxes and customs duties, which, according to one estimate, could amount to as much as 45 percent of the total income the Chadian government derives from the project. Perhaps more remarkable, the law covers only income from the first three fields to come onstream in the Doba basin—Miandoum, Kome, and Bolobo. In May 2005 Exxon-Mobil began producing oil from the Nya satellite field, another came onstream in March 2006, and three more were in the conceptual phase as of late 2006. There are also large concession blocks in northern and central Chad, where Canadian, Swiss, and Chinese companies have been drilling exploratory wells, with some success. None of these would be covered by Law 001. Initially ExxonMobil indicated that, as an act of good faith, it would consider its satellite fields to be covered by the original agreement and continue depositing sales receipts in the London Citibank account, but the company was quickly pressured by the Chad government to pay directly into Chadian banks.

Additionally, the law is vague about what constitutes priority sector or regional spending. What, for instance, does "health and social services" really mean? And what exactly is considered the "oil-producing region" for the purposes of receiving the special 5 percent allocation? There have been squabbles over definitions, and the government has sometimes interpreted terms in questionable ways. In 2004, for example, more than half the money made available for priority sectors was spent on building two roads, while education received only 5.1 percent and health only 3.3 percent of the oil revenue.

Critics also contend that the law has not succeeded in helping Chad avoid the *rentier* trap. In recent years, the state's tax receipts have plummeted. In part this is because corrupt officials have assumed that, with all the oil money sloshing around, less attention is being paid to traditional revenues and they are less likely to be caught stealing from the treasury. And, in fact, there is less emphasis on traditional revenue. Father Berilengar lamented, "We don't talk about tax and other forms of revenue anymore. It's all oil money. If there's any kind of disagreement, or if functionaries haven't been paid on time, people just automatically say, 'Well, where's the oil money?' They claim we're not letting them spend the oil money, which is just an excuse. The country lived without oil before, and those resources are still there. But the government just blames the *Collège* and Law 001. And we're stuck in the middle." Indeed, many believe Law 001 has made the oil money too bureaucratically entangled to be put to good use. "The money is just sitting there," Father Berilengar said to me with a frustrated shake of the head. "The bureaucrats are afraid of too many controls." Functionaries were more likely to spend small sums on cars and office furniture, which didn't require a lot of signatures, than to tackle the process of getting larger sums approved by the *Collège* for real development projects.

The need for such projects has been felt most keenly by those living closest to the Doba basin oil fields, in southern Chad. Under Law 001, the region is supposed to receive 4.5 percent of profits from oil exploration as a special supplement to be distributed by local authorities, in recognition of the disruption to the lives of local people. Critics have complained that this figure is too low—compare it with the 13 percent set aside in Nigeria, which is seen as not nearly enough by activists there—but the more-pressing concern is that the allocation applies only to the first five years of the project and can thereafter be canceled by presidential decree.

Disagreements between multinational corporations and destitute African villages often turn into a needlessly polarized ideological battle between proponents and opponents of globalization and free-market capitalism, or into an oversimplified David and Goliath tale. I wanted to see for myself the situation around the Doba basin and whether critics were justified in heaping so much blame onto Exxon-Mobil. Getting there from N'Djaména was going to be a challenge, though. In 2005 Chad's national airline, Toumaï Air Chad, was down to one functioning plane, a battered 737 servicing six African destinations and one domestic airport in the east of the country, as well as the annual pilgrimage to Mecca. Until such time as it was able to purchase a second aircraft, Toumaï regretted that it would not be providing service to southern or northern Chad.

I asked about how I might make the journey by land, but received bewildered looks and was sternly warned that it would be a rough and dusty 200-mile trek across the blazing heat of the Sahel, and very much not for the faint of heart.

None of this was a problem for ExxonMobil staff, of course, because the company had its own airport and chartered a fleet of planes making regular flights between N'Djaména and Doba. In the early days of the project, when it was still being hailed as a "model" for African oil exploration, Exxon had been happy to fly journalists south, and had bent over backward to set up tours and meetings with local managers. But this was 2005, and ExxonMobil had been burned by a slew of negative stories in the international press. So, when I approached the company six months in advance, I was told by its Houston PR department that arranging a flight would not be possible. Even if I somehow made it to Doba under my own auspices, in fact, it would not be possible to have a tour of the company's project in southern Chad. Nor was I allowed to speak to Exxon staff at any point while I was in the country, not even off the record. Any questions I had would be answered by Houston.

Given the cost of hiring a car and driver for the two-day round-trip south—at least $200 a day—public transport rapidly emerged as my only option. At the crack of dawn one Thursday, I watched as my suitcase was lifted to the roof of a beaten-up old Land Cruiser and steadily squashed under a small mountain of accumulating bags and boxes and threadbare trunks. Off to one side, a young man unscrewed the vehicle's fuel cap and stuck a piece of rubber tubing into the tank, to the other end of which he stuck a small plastic funnel. Out of nowhere, several glass jugs appeared, filled with gasoline, which the man steadily poured into the funnel, taking great care not to spill any.

It was as powerful an image as one could possibly ask for. Chad may have recently joined the ranks of the world's oil-producing countries, but the country still lacks a downstream oil sector, and its citizens have yet to see what an actual gas station looks like. There is no refinery for Doba crude to be sent to, so every last drop of Chad's crude goes straight into the ExxonMobil pipeline and straight onto supertankers parked off the Cameroonian coast. There are few cars in Chad, but those that exist (almost all of them taxis or official vehicles) operate not on Chadian oil but on Nigerian gasoline. The refined product is driven—often smuggled—across the border, and sold from glass jars in shaded spots along the side of the road that look like little more than American-style lemonade stands.

Inside, the Land Cruiser had been converted into a sort of cattle truck, with two hard wooden benches running along its length. Ten people had already squeezed in, along with more belongings, and were looking intensely uncomfortable in the scorching early-morning heat. I plumped for a $25 "first-class" ticket, thinking that sitting in the passenger seat and facing forward would make the ten-hour journey more pleasant. What I had not been told was that a first-class ticket entitled me to only *half* the passenger seat.

It was just as well, then, that the vehicle broke down at least eight times during the journey. (I lost count after the seventh.) The passenger seat was tilted so far forward that my neighbor and I had our arms pressed against the dashboard for the duration of the journey, and every time the driver pulled over to fiddle with the fan belt, it was a welcome chance to step out into the blessed relief of the 120-degree heat and walk around among the camels and stray goats and round mud huts, silently cursing ExxonMobil until feeling returned to my arms and legs. On the way back, two days later, I treated myself to *both* first-class seats. After all, it was my birthday.

◄◄━►►

"The government ignored all the negatives. They told the population that oil would be a paradise, that it would solve all of their problems. But we saw the experience of Nigeria and others and wanted to ensure that the population was informed about the reality." It was my understanding that I was speaking to Nadji Nelambaye, coordinator of the local coalition of NGOs, but since it was pitch-black and I was using my lone candle to help me take notes, I could have been talking to anybody.

Like more than 98 percent of Chadians, the residents of Moundou have no access to electricity. In one of the extraordinary ironies of Chad's oil boom, this energy-rich country's dilapidated grid provides, at the best of times, a mere 20 megawatts of electricity. The world's newest oil producer literally cannot keep the lights on. Moundou, like most of Chad, spends its nights in total darkness, save for the flickers of gas lamps and candles, or the headlights of passing motorcycles.

Meanwhile, in nearby Kome, the twenty-five-mile-wide Exxon-Mobil facility lights up the night sky for miles around thanks to its state-of-the-art 120-megawatt generating plant. Not only does the ExxonMobil compound produce six times as much electricity as the entire Republic of Chad, it likely produces as much as the entire Sahel. So bright is the light from Kome, and so dark is everything around it, in fact, that the facility is visible from outer space.

Moundou is very much *not* Chad's answer to Port Harcourt. ExxonMobil has confined its operations to the fenced compound at Kome, some fifty miles away, and Moundou has continued to languish as a dusty backwater, where even the town's three hotels no longer bother to repair their broken generators. Moundou's only claim to fame is that it is home to Chad's national brewery, where the bottles of bland Chari beer famously continued to roll off the assembly line throughout the darkest and most savage years of civil war. According to the World Bank, Moundou, a town of 96,000 people, has only two doctors.

Given the oppressive heat and the lack of light, Nadji suggested we reconvene early the next morning for a trip to Kome. But when morning came, I wondered how I would find Nadji, since I still didn't know what he looked like. Fortunately for me, in Moundou I stuck out like a black man at a Merle Haggard concert, and Nadji quickly found me wandering a deserted street at the edge of town, looking for his office.

On the drive to Kome, Nadji rattled off some of the problems that ExxonMobil's presence was believed to have caused. The coalition had done a study showing that eleven primary schools had closed, thanks to teachers leaving to find more lucrative—if temporary—jobs with ExxonMobil. Worse, many girls had given up on school entirely to work outside the oilfields as prostitutes, and the rate of AIDS infection was increasing. Young men, meanwhile, had abandoned their fields to look for work at ExxonMobil, resulting in a decline in agricultural productivity and an accompanying rise in the local price of millet—a situation exacerbated by the increase in demand for grain from people working for ExxonMobil. The government had not stepped in to regulate prices and local people had suffered hardship.

As we drove along the red-dirt track, enormous construction trucks loaded with Filipino laborers passed by every few minutes, kicking up blinding clouds of dust and diesel exhaust. Nadji didn't miss a beat. He explained that the coalition had tracked an increase in respiratory illness among the local population since the project began and had pressured ExxonMobil to address the problem. Exxon, he said, had refused to pave the road, claiming that was the government's job, and had instead watered the road to keep the dust down. In the desert heat, though, the water evaporated quickly. Within hours, the dust was back.

Nadji continued describing the social disruptions the coalition was tracking. During the construction phase of the project, he said, Exxon-Mobil subcontractors had trained locals to act as paramilitaries. Since the construction ended, however, most of the locals had gone back to their villages and, unaccustomed to making ends meet without the generous salaries paid by the contractors, had put their newfound skills to use in aggressive acts of criminality and banditry. Divorce rates had also gone up, thanks to displaced farmers spending their compensation packages on prostitutes. "If you take a poor, rural man who has never seen more than $5 or $6 in his hand and you give him $2,000 in compensation, he is likely to spend it on beer and girls." With less land to go around since ExxonMobil moved into the area, farmers and animal husbandmen had also been driven into nasty conflicts.

The list of complaints went on and on.

After an hour, we arrived outside ExxonMobil's Kome operating base, and I immediately saw why the company had become reluctant to bring journalists here for show-and-tell (as well as why Nadji was so keen for me to see it). On one side of the road, surrounded by a high perimeter fence, was the base—an ultramodern, air-conditioned facility with its own airport, powered by four electric turbines and protected by armed guards. A sign next to one of the buildings welcomed visitors to Kome, which it declared, in a chunky typeface reminiscent of Midwestern roadside advertising, to be "Home of the World's Greatest Drilling Team." On the other side of the road was a stinking, ramshackle slum, which a far-more-modest road sign identified as "Atan."

Ten years ago, neither Atan nor the Kome base existed. The area had been home to a few hundred pastoralists living in clusters of round mud huts. But when Exxon began building Kome, word got out that the company would need a few hundred laborers, and people poured in from miles around. They stood for hours and days outside the perimeter fence, in the hopes of snapping up even a temporary job. Days turned into weeks and months, and a small squatter camp grew up outside Kome. The presence of large numbers of young men attracted girls, who had heard there was a good living to be made as prostitutes. Before long, the girls were coming from neighboring Nigeria, Cameroon, and the Central African Republic, and even Ghana. As the squatter camp grew, its residents nicknamed it the *Quartier Attend*, which roughly translates as "Waitsville" or "Waiting Town." *The place where people wait.*

With its transient population of young laborers and girls from all over West Africa, *Quartier Attend* developed a reputation as a place of loose morals, and people began referring to it jokingly as *Quartier Satan*, or "Devil-town." (In French, *Satan* rhymes with *attend*.) At its height, it was home to as many as 17,000 people, many of them from as far away as Morocco and the Philippines. Some worked as drivers or security guards for ExxonMobil, but others were just attracted by the dynamic economy. Families began settling in Attend, and small primary schools were set up, along with a mosque and a church, and even a small cinema. The village elected a chief and got itself officially recognized by the government as a town on the map of Chad. And, in a touching display of civic pride, it asked to be called Atan, which, although pronounced in the same way as Attend, lacked the baggage of the town's dubious beginnings, and almost looked like an authentic, phonetically spelled African name.

Despite the veneer of respectability, Atan is an enormous festering embarrassment for ExxonMobil—a living, breathing metaphor for the failure of the Doba drilling operation to bring meaningful development to the people of Chad. On one side of the road, Exxon employees enjoy modern rooms, complete with private bathrooms, DVD players, and Internet connections. They are cared for in a modern clinic and can unwind on basketball courts, to which there will soon be added a tennis court and swimming pool. On the other side of the road, in a makeshift camp, some 10,000 people make do without clean running water.

When I began taking snapshots, Nadji quickly made me stuff my camera into my bag, warning me that I would get my film confiscated if I wasn't careful. ExxonMobil, he said, paid plainclothes *"vigilants"* to stop anyone taking pictures, even if the camera was not pointed at the Kome base. So we made do with wandering the streets of Atan, admiring the improvised shops and stalls selling everything from cigarettes to fried meat to a vicious home-brew called *bili-bili*.

Despite its schools and houses of worship, Atan has not entirely shed its sin-city image. Next to each other along the road, and within easy access of the base, are two "nightclubs." One, called Phoenix, is favored by the French workers from Kome, while the other, La Maison Blanche Number One (White House Number One), is staffed by English-speaking girls from Nigeria and Ghana and caters mostly to an American clientele. We stepped out of the sun and into the Phoenix and found it mostly empty. After all, it was still early in the morning. I noticed a passageway that led to a semiprivate spot behind the nightclub, where the girls would take their tricks for sex, and Nadji told me he had last been here with a French television crew who had come late at night and filmed illicitly for a documentary. Although Atan was a public space where, with the appropriate permits, any journalist ought to be allowed to film, ExxonMobil's influence in Chad meant that the extraordinary visual narrative of this town's coexistence with the Kome base could never be documented properly for a Western audience.

Nadji began to look a bit nervous and suggested we leave before our presence drew too much interest. We drove down the road a few miles to Ngalaba, one of three traditional villages that had become known as "*villages enclavés*"—enclaved villages. Ngalaba, along with nearby Maikeuri and Bendoh, was cut off from some of its traditional grazing lands by power lines and feeder pipelines when the Doba project got under way, and villagers say their livelihoods have been destroyed. ExxonMobil insists that its facilities pose no danger to the villagers and that they have been compensated for the loss of their arable land.

Ngalaba is a village of 1,125 people, led by a traditional ruler named Tamro, a quiet and thoughtful man in his late thirties or early forties, who at first seemed hesitant to talk to us. Speaking in the local Ngambaye language, which Nadji translated into French for me, Chief Tamro looked into the distance and admitted that he was "worried." He had noticed that the mangoes had failed to thrive this year, and wondered if it was because of the gas flare from a nearby well. He complained that ExxonMobil had left some of its exploratory wells unplugged, and that village livestock had fallen in. "We lost many animals that way," he said, before adding that Exxon had responded to their complaints about dust by coating the dirt road with molasses, which is toxic to goats and cattle. "I am very worried," he repeated, so quietly that we could barely hear him. "Honestly, I would rather they just found us another piece of land and we could all go there and leave the village." The men who had gathered around us looked genuinely saddened and disappointed by what they heard their chief saying. "We need to start over. There is no security here."

"This is our land," a dark and round-faced young man named Judé piped up. "We've seen no benefit from it. We lost our land and have received nothing for it. At first they said they were going to build hospitals and dispensaries here. But they've done none of that." Exxon, Chief Tamro explained, had offered the village its choice of five options: a school, a well, a granary, one kilometer of paved road, or a marketplace. The villagers chose the school, understanding that it would house six grades, but ExxonMobil built them a two-room schoolhouse instead. "Let me ask you something, sir." The chief tried to contain his frustration. "If I take something from you, should I then come and dictate the terms of my compensation to you for the loss? Surely it is for me to apologize and ask you what I can do to make it up to you."

He pointed out a tiny, windowless concrete shed that stood out among the round straw and mud huts. "They told me that they spent 30 million francs [about $60,000] on that house, and that their workers were going to live in it. In the end I had to break the door down so I could sleep in it myself." The men all shook their heads. "I ask you," the chief said, "does that look like a 30 million–franc house? You know how much I could have done for this village with 30 million francs?"

Back in N'Djaména, I sent an e-mail to Houston, asking Exxon-Mobil to respond to everything I had seen and heard around Kome — the dust, the AIDS and the prostitution, the *villages enclavés,* the goats falling into uncovered wells, the two-room schoolhouse, and the $60,000 shed. "I would suggest that as a first step, you familiarize yourself with the basic facts on what has occurred," came the reply. "I can help by sending you a CD containing our Quarterly Reports." The CD never arrived, but as of this writing, the reports are still available on the ExxonMobil Web site at http://www.exxonmobil .com/Chad/Library/Reports/Chad_QuarterlyReports.asp.

All 1,200 pages of them. I suggest that you familiarize yourself with them.

Having gotten nowhere with ExxonMobil, I decided to try my luck with the World Bank. As a publicly funded international organization, it would have to have a little more time for the press. And sure enough, its resident representative in N'Djaména, Noël Tshiani, an enormously charming Congolese man, welcomed me into his frosty office on one of my last evenings in Chad, and apologized for keeping me waiting outside. I would have been happy to sit for hours in the World Bank HQ, since it was the first time I'd been able to stop sweating in ten days, but I smiled and said it was no problem. In front of Tshiani sat a bowl of preserved peaches—a lunch he had not had the chance to eat. It seemed appropriate that Tshiani was run off his feet this late into the evening. He had presided over a particularly turbulent time in relations between the Chadian government and the Bank and, though neither of us knew it then, things were about to get worse.

Tshiani had arrived in N'Djaména in October 2004, and, "almost within the first five minutes," as he described it, he had received a phone call from President Déby accusing Exxon of "cheating." At issue was the fact that the government's receipts from Doba crude had been fixed at $25 a barrel, but (thanks in part to the antics of Dokubo Asari across the border in Nigeria) oil had been selling for upward of $50 a barrel on the international market. The Chadians felt they were being ripped off, and President Déby had even issued a statement provocatively titled "A Consortium's Fraud and Secrecy in Exploiting Doba's Crude." In actual fact, Exxon was well within its rights, as Doba crude is of very poor quality and costs a fortune to transport to Kribi.

But what was even more irritating to the Chadian government was that even this $25 a barrel was largely out of its reach. The original agreement gave Chad only 12.5 percent of its own revenue, most of which was contractually obliged to the Future Generations Fund or to priority sector and regional spending. In other words, as the price of crude was soaring past $50 a barrel, the government was getting barely $3 a barrel, most of which was beyond its grasp. By the time all its obligations and conditions were met, Chad's government was getting less than fifty cents a barrel in real income to spend as it wished. Tshiani was hauled into the National Assembly to "explain." "I expected the whole thing to take an hour," he told me, "but it ended up taking all day."

I asked Tshiani about the "two-speed problem" and wondered whether the Bank was now prepared to concede that it had been too optimistic about the speed of Chad's capacity building. "I agree that this country has very weak capacity, across all sectors," he replied. "But when the pipeline was being built, we had two choices. Either we don't start building the pipeline and wait until all the capacity is in place, or we adopt a system of parallel tracks. We chose parallel tracks. The only problem we had was that the pipeline was built one year ahead of schedule. If we were starting again, I would say we should have started the capacity building earlier. But the construction took place much faster than we thought." This sounded like an admission that the NGOs had been right when they pleaded with the World Bank to slow down the project. But Tshiani believed many of the weaknesses in Law 001 would be ironed out over time. The mandate and resources of the *Collège* would be extended, he said, and the government had assured him that the "principles" behind the law would apply to the five new satellite fields. "I truly believe the government means well," he stressed.

Perhaps the government did "mean well," but we would never find out. Soon after I left Chad in April 2005, the situation in the country began unraveling. That autumn President Déby found himself faced with increasing dissatisfaction from within his already-limited power base. At issue was his autocratic style and his crude push to amend the national constitution so he could stand for a third term as president—something he had earlier promised he would not do*— and also the simmering rebellion in the western region of Darfur in neighboring Sudan, which pitted ethnic Zaghawas against Janjawid militias backed by the government in Khartoum. Déby was in an awkward position. The Zaghawa ruling circle in Chad wanted him to do more for his kinsmen being slaughtered in Darfur, but Déby could not forget that in 1990, it was the Sudanese government who provided him with a rear base from which to invade eastern Chad and depose Hissène Habré.

By late 2005 Déby's own uncles, the former cabinet ministers Tom and Timane Erdimi, had started their own rebel movement. In mid-November an attack on the military barracks in N'Djaména was reported and shots were heard. Later that month, a group of soldiers, led by a fresh-faced thirty-one-year-old just back from completing an electrical-engineering degree in Ottawa, deserted the army and set up camp in the east of the country, calling themselves the Platform for Change, Unity, and Democracy. The Erdimi twins quickly joined their ranks. A few days later, Déby dissolved his presidential guard, and dismissed key Zaghawas from positions of power.

With the loyalty of the Zaghawa patronage network and the generals no longer secure, Déby had no clear power base. Zaghawas make up only 2 percent of Chad's population, so there would be no grassroots uprising in his favor. Déby desperately needed to get his hands on oil revenue in order to bolster his domestic position. From Déby's perspective, the unity of Chad as a nation was at stake, and this was no time for the World Bank to sermonize about priority sectors. In December, therefore, he had the rubber-stamp National Assembly amend

---

*During his previous election campaign in 2001, Déby told a French newspaper, "I will not stand as a candidate in 2006. I will not change the constitution—even if I have a 100 percent majority."

Law 001, extending the definition of "priority sectors" to include not just schools and hospitals and roads, but also, in a classic act of legerdemain, "domestic security." In other words, Chad's oil money could now be used to buy weapons for fighting what increasingly appeared to be a rebellion.

The World Bank went ballistic. Its new president, Paul Wolfowitz, freshly reshuffled out of the Pentagon after his drive to invade Iraq, held an angry two-hour phone call with Déby on January 7, 2006. The Bank then announced it would be freezing Chad's accounts in London and suspending the entirety of its $124 million debt-relief package in Chad. Déby looked increasingly isolated, but he charged ahead, calling elections for May 2006. Credible opposition parties were disqualified from standing, and it became conventional wisdom that by announcing a date for his sham reelection Déby had succeeded only in setting a timetable for his own violent overthrow. The rebels in the east gathered force and, on April 13, 2006, launched an attack on N'Djaména that was put down following heavy fighting that included an aerial bombardment of the capital by forces loyal to Déby. The president survived the rebellion, but most observers chalked this up at least in part to the incompetence of the rebels: When they arrived in N'Djaména after weeks of fighting their way across the vast eastern stretches of the country, they were seen asking bewildered residents for directions to the presidential palace.

Chez nous, le pouvoir vient toujours de l'Est, goes a popular saying in Chad—"Power comes from the East"—and 2005–2006 was proving no exception. Both Déby and his predecessor Habré had launched their coups from Sudan, and now it seemed Déby would die by the sword he had lived by. In December 2005 he angrily accused Sudan of supporting Chadian rebels in Darfur, which the Sudanese denied, accusing Chad of harboring Darfurian Zaghawa rebels on the Chadian side of the (increasingly fictitious) border. The two countries declared a "state of belligerence" and in April 2006 broke off ties completely. Déby was now almost entirely on his own. His traditional protector, France, flew in 150 soldiers to add to its permanent presence of 1,200 military personnel and provided Déby with aerial reconnaissance photos of rebel bases, but respectfully declined to do more.

Even in 2005, N'Djaména had felt like a city on the edge of a nervous breakdown. Military convoys bristling with cocked machine guns sped down the dusty streets, and the managers of my hotel on the relatively safe avenue Charles de Gaulle would not let me venture two doors down to the Internet café without a security guard accompanying me. It struck me as an excessive precaution until the day I saw a pair of French soldiers being attacked and mugged in broad daylight in front of the hotel.

Insecurity sometimes seems to be the way of life in the Sahel, where nothing—not even the rains—is guaranteed. At press time, ongoing violence in the western Darfur region of Sudan had begun to spill into eastern Chad, a situation that had always threatened to manifest itself one way or another. For years Déby had provided quiet support to disgruntled Zaghawa kinsmen on the Sudanese side of the border and, by late 2005, it was obvious that Khartoum had decided to retaliate by arming Chadian rebels in Sudan.

As for Déby, he has surprised everyone simply by surviving and, by summer 2006, even appeared to be consolidating his position. The World Bank was forced to capitulate, and Law 001 was renegotiated in Déby's favor. In August relations with Sudan were reestablished and ties with Taiwan were abruptly severed in favor of a robust new relationship with Beijing. And the ultimate coup de grâce came on August 26, when Déby went on the radio to accuse Chevron and Petronas of fudging their taxes, in violation of their agreements with Chad. Three cabinet ministers, including the oil minister, Mahmat Hassan Nasser, were being relieved of their duties, the president declared, and Chevron and Petronas had twenty-four hours to pack their bags and leave the country.

As eagerly as it had been anticipated, the Chadian "model" had unraveled. What had begun in the 1990s as a beacon to future African oil producers, an example that a petroleum bonanza could serve as an engine for development and prosperity, and a lifeline out of poverty for millions of struggling citizens, had, within a few short years, turned into yet another farrago of embarrassing headlines from a dusty corner of Africa where white men had found oil. By early 2006 it was being widely reported that Déby could no longer rely on the security of his Presidential Guard, and was spending much of his time being spirited around at high speed from bunker to bunker in an armor-plated Hummer, while body doubles and look-alikes were being paid to be seen boarding the presidential jet. The Bank was embarrassed, the people of Chad were still destitute, and somewhere in Texas the world's largest oil company was chalking up yet another year of record profits. As for Noël Tshiani, the World Bank rep, he was suspended at the end of 2005 following allegations of sexual harassment. Nothing, it seemed, was going to go right with this picture.

As 2006 drew to a close, another rebellion appeared to be brewing in the east of the country. Government forces fought vicious battles near the Sudanese border, as rebel columns advanced yet again toward the capital, taking two key towns along the way. On October 29, the head of Chad's armed forces was killed during a particularly fierce battle with rebels.

◄◄─►►

In the end, Déby never followed through on his demand that Chevron and Petronas leave the country. But then, he never really had to. For most analysts, it was an empty threat, designed to rattle the World Bank and remind Paris and Washington (as well as internal enemies) who was really in control in N'Djaména. Coming, as it did, just a few days after the abrupt switch in diplomatic relations from Taiwan to China, the action against the oil companies was seen by many as a clear sign that Beijing was seriously interested in becoming involved with Chad's oil boom, and that Déby was more than willing to entertain the possibility of falling under a Chinese sphere of influence, even if it meant sacrificing existing arrangements with Western companies.

Déby's ability to strike fear into the international business community simply by engaging in a little aggressive flirting with Chinese politicians was not an isolated event. By late 2006, in fact, the invisible hand of China was being seen all over Africa, and it was increasingly anything but invisible.

CHAPTER 7

THE CHINESE ARE COMING! . . . BUT WHO ISN'T?

IN 1985 NEWSPAPERS IN BEIJING ran a front-page story about an exciting technological innovation that had become available to ordinary Chinese citizens, which they touted as the latest indication of the soaring living standard in the countryside. In a distant rural province, a farmer had become the first in China to buy his own truck. The papers showed pictures of the beaming comrade at the wheel, an icon of pride and progress.

Twenty years later, there were some 20 million motor vehicles in China. The count is expected to hit 56 million by 2010 and 140 million by 2020. The image of the People's Republic most of us grew up with—that of great swarms of ashen-faced cyclists trundling along on government-issued Flying Pigeons—is a thing of the past. Modern Chinese cities now look increasingly like Hong Kong, with streams of shiny Volkswagens and Mitsubishis speeding along newly built overpasses and into the inevitable traffic jams.

These cars are only the most visible sign of China's booming economy, with both factories and a burgeoning educated middle class demanding ever more state-of-the-art technology to go about their business and their lives. The country is increasingly dependent on affordable energy, and increasingly less able to meet that demand from its domestic sources. For years, China's predominantly rural economy and top-down urban-transport policies meant it had more than enough oil to meet its needs. In the mid-1980s, when our farmer friend was grinning for photographers in front of his new truck,

China was the second largest *exporter* of crude oil in Asia. But in 1996 China officially crossed the line from being a net exporter to being a net importer of oil and, by 2005, it had overtaken Japan to become the world's second-largest importer—behind only the United States.

One-fifth of all human beings live in China. News that a country that size can no longer meet its energy needs is bound to make petroleum traders nervous and contribute to a sustained period of high global energy prices—as has been the case over the past several years and as is likely to be the case for several more years. Forecasters expect China to have to source 60 percent of its energy needs from abroad by 2020. Even for a much smaller country, this would pose a challenge to the global energy market, but in this case Beijing will need to import 10 to 15 million barrels of oil a day—more than double the current output of Saudi Arabia. Or, put another way, more than the entire output of the African continent.

Where is all this oil going to come from? For Beijing the answer initially seemed to be neighboring Asian countries, such as Indonesia or Brunei. But as the scale of China's impending economic miracle became apparent, China quickly realized that more-distant shores would have to be sought out. One of the most important of these would be Africa. Unlike in Western countries, where oil companies are usually independent from governments, in China virtually all petroleum exploration is undertaken by state-owned firms, which means there is a direct relationship between Beijing's foreign policy and the commercial affairs of its oil industry. It has become fashionable for Americans to point out the cozy relationship between their elected politicians and the U.S. oil industry, but in China there is no meaningful way to talk about government and industry as if they were separate entities. For China's politicians, energy security is an explicit goal of the nation's foreign policy, and Africa is a region of increasing strategic importance. In the 1990s, aware that China had to catch up with Western countries whose multinationals had operated in Africa for decades, Chinese officials made the entry of Chinese oil companies into Africa a top priority. In 1997 African oil accounted for 17 percent of China's imports. By 2004 that figure had climbed to 28.7 percent, and it will probably continue to climb in coming years—making Africa even more important to China, from an energy-security perspective, than it is to the United States.

As a result, Beijing has not been shy about stepping up its political and economic activities across the African continent. Many in Washington have nervously pointed out that, as the United States gets bogged down in imperial adventures, China has steadily raised its profile in parts of the world that are lower down Washington's priority list. Nowhere has this been more true than in Africa, where China has been launching partnerships and promoting direct investment by its companies.

Taking advantage of traditional diplomatic friendships as well as the fairly wide-open playing field that still exists for oil exploration on the continent—in other words, the same factors that attracted Western companies to Africa—the Chinese have been able to snap up lucrative new exploration acreage. By and large, their strategy has been to offer sizable inducements in the form of cash loans or development projects, which—in stark contrast to Western forms of debt relief—come with no real strings attached, no endless sermonizing about fiscal responsibility, and no micromanaging of government spending. While the World Bank was earnestly trying to channel Chad's oil wealth into health and education projects, for example, China was happy simply to give the Angolan government $2 billion to build roads and airports—in exchange for promising offshore acreage as well as contracts for the purchase of crude from the state oil company Sonangol. Largely as a result of such checkbook diplomacy, Angola has become the largest and most important source of oil for China, overtaking Saudi Arabia. In 2004 China even overtook the United States as Angola's biggest customer for crude.

Chinese bilateral trade with Angola jumped 113 percent in 2004, to a whopping $4.9 billion. But it is not only in oil-rich states that

China is increasing its commercial presence. Overall Chinese trade with Africa trebled between 2000 and 2005, to nearly $50 billion, and is expected to hit $100 billion by 2010. (In 1989 it was not even $1 billion.) In 2006 China overtook Britain as Africa's third-largest trading partner, and has number-two France well within its sights.

Even beyond these statistics, the scale and the ferocity of China's entry into Africa has been breathtaking. China has started construction on a new railway in Nigeria and a new port for Gabon, has paved most of the roads in Rwanda, and is building roads, bridges, power stations, schools, and cellular-phone networks in at least a dozen African nations. At any given time, the China Road and Bridge Corporation alone is likely to be engaged in five hundred projects throughout Africa. In tiny Lesotho, nearly half the supermarkets are owned and run by Chinese, who also operate textile factories in the country. Mauritius, home to many Chinese-owned textile factories, added Chinese language to the national school curriculum in 2004.

The Chinese have not hesitated to back up their commercial activities in Africa with political and diplomatic efforts. In 2003 Prime Minister Wen Jiabao toured several African oil-producing countries in the company of senior Chinese oil executives, while President Hu Jintao visited Algeria, Egypt, and Gabon. In June 2006 Wen made another African trip, to seven countries including Angola and Congo-Brazzaville. Chinese embassies have been opened or expanded and consular representation upgraded, particularly in such countries as Ethiopia, where Beijing expects to be drilling for oil in coming years, and the few African countries that continue to recognize the independence of Taiwan have been gently persuaded of the merits of a one-China policy. Idriss Déby's abrupt rebuff to Taipei in August 2006, paving the way for CNPC to drill and produce in northern Chad, is only the most recent example.

Beijing has been particularly good at invoking the "spirit of Bandung" in recent years—a reference to the 1955 conference in Bandung, Indonesia, that established the Non-Aligned Movement. The goal of Bandung was to bring together developing countries that wanted to take a neutral position in the Cold War but feared being left on the sidelines while the superpowers doled out aid packages to their allies. The Non-Aligned Movement is largely irrelevant today, but for many years it served the Chinese as an important source of influence in Africa, with thousands of Chinese doctors being sent to Africa between the 1950s and the 1970s and thousands more African students completing their educations in China. More recently, the Chinese have been happy to make use of the networks they built on the continent during those years and to remind African leaders which superpower stood by them through thick and thin and never criticized their internal policies. China pointedly used the occasion of the fiftieth anniversary of Bandung, April 23, 2005, to launch its New Asian African Strategic Partnership (NAASP)—a more robust version of the China Africa Cooperation Forum (CACF) it set up in 2000 to promote trade and investment with forty-four African countries. The launch announcement for NAASP was even made in Bandung.

But the most spectacular demonstration of China's renewed commitment to Africa came in November 2006, when more than forty African heads of state, along with 1,500 other delegates, gathered in Beijing for a special "China-Africa summit" organized by the CACF. The event, which was described by its hosts as "a new landmark" in Chinese foreign policy, was the largest and most-high-level gathering of world leaders in Beijing since the founding of the People's Republic. The assembled heads of state alone represented a quarter of the votes at the United Nations, and (along with their hosts) a third of the planet's population.

In terms of sheer scale and ambition, the summit dwarfed anything that Britain, France, or the United States had achieved—or even attempted—for Africa in the past. One million Chinese citizens were mobilized to provide security, transport, and entertainment. The Great Hall of the People hosted a dazzling display of acrobatics and African drumming, and the normally chaotic, smog-choked streets of

Beijing were cleared of traffic for three days, as hundreds of thousands of drivers were ordered to stay at home. Nearly every billboard and wall in sight was covered with giant pictures of sprawling savannahs, giraffes, elephants, and half-naked tribesmen, along with captions proclaiming solidarity between the peoples of China and Africa. Hotel rooms were fitted with African furnishings, and hotel staff were given lessons in Swahili and French. At the conference itself, a whopping $5 billion in new loans and credits for Africa were announced by the Chinese hosts, who also pledged to train 15,000 African professionals and establish a development fund for building schools and hospitals across the continent.

Part of the reason China has been able to expand its presence in African oil exploration so rapidly has been its ability to take a long-term approach. Unlike Shell or ExxonMobil, CNPC is backed by the state and doesn't have to worry as much about volatilities in the price of oil, or the instability of the African political environment. In the absence of demands from rambunctious shareholders, CNPC can snap up commercially or politically unappealing acreage and simply wait out the difficulties. As a result, the Chinese have been most clearly visible in Africa's marginal or declining oilfields, where the less-risk-tolerant majors cannot justify the cost of being involved, or in countries, such as Sudan, where human-rights concerns or even sanctions keep Western companies out entirely. Analysts believe the "master strategy," if there is one, is to pick up quick drilling expertise and experience in less-desirable spots and be able to compete with Western majors when new lucrative licenses become available in more proven terrain.

It's a strategy that suits the Africans as well as it suits the Chinese. Gabon, for example, has been written off as a "declining" oil power, with Western oil giants Total and Shell scaling back their activities in recent years. But in 2004 China and Gabon signed a series of deals under which China agreed to look into the possibility of building a second refinery for Gabon and in return received a couple of "marginal" exploration blocks, along with the promise of 20,000 barrels of Gabonese crude a day. Thanks to China, Gabon will be able to squeeze every last drop out of its oil reserves, and thanks to Gabon, China will pick up valuable experience drilling for oil in Africa.

In the same year, a similar deal was reached whereby CNPC was granted exploration rights in the neglected Lake Chad basin of northern Nigeria and, after much arm-twisting by Nigerian officials, promised to help revive the country's hapless downstream sector. Western analysts mocked the deal, pointing out that Nigeria's three antiquated refineries suffer regular breakdowns as well as frequent sabotage by crude-oil bandits, and that entrenched interests in Nigeria prefer to see refined fuels imported. "The Chinese are more than welcome to Nigeria's downstream," chuckled a U.S. official when I raised the issue.

Nevertheless, others believe the Chinese are shrewdly currying favor in Abuja, with the ultimate aim of shoehorning their way into Nigeria's far-more-profitable upstream sector. They point out that without the cumbersome HR departments, employee-benefit packages, and slick PR machines of Western companies, the Chinese have proven much better at getting their heads down and their hands dirty, and going about the business of oil production, even under the most adverse circumstances—a fact confirmed by their ability to pump oil out of southern Sudan during the worst ravages of the Sudanese civil war. As one rather hard-boiled American I met in the Delta put it, "The Chinese come to Nigeria, and they think they've died and gone to heaven."

◄◄•►►

Anyone who has been to both Nigeria and southern Sudan will agree there is a great deal of truth to this observation. However chaotic and unstable the Niger Delta may have become, it cannot compare to southern Sudan, which must rank as one of the most inhospitable commercial operating environments in the world. From 1956 to 1972, and again from 1984 to 2005, the southern states of Sudan were the scene of one of Africa's most brutal and intractable civil wars, as the mostly black animist and Christianized South fought against various Arab- and Muslim-dominated governments in Khartoum. Some 1.5 million people were killed during the conflict, many by disease and hunger. Leaving aside the war's pernicious effects, the vast region has been neglected by distant colonial rulers for well over a century—first by the Ottoman Turks, then by the British-Egyptian condominium that ruled Sudan from 1898 to 1956, and finally by the authorities in independent Khartoum. As a result, even as the final signatures were being put on the peace accord that brought the war to a close in January 2005, southern Sudan was quite literally living in the Stone Age.

Three months after the Comprehensive Peace Agreement was signed in Nairobi, I traveled to southern Sudan. At the time, analysts were talking about 600 million barrels of reserves in southern Sudan, but many now believe the figure is closer to 1 billion. Already, the Chinese, along with companies from Sweden, Pakistan, Malaysia, India, and elsewhere, were pumping 400,000 barrels per day out of the country, and were expected to top 700,000 barrels per day by the end of 2007. Whether Sudan is a sub-Saharan or North African country is a matter of perspective, but in sub-Saharan Africa, that kind of production would catapult Sudan past Gabon, Equatorial Guinea, and Congo Brazzaville, and put it behind only Nigeria and Angola.

Even in 2005, getting to southern Sudan was a tricky business. The government in Khartoum exercised only patchy control over the region, and the separatist Sudan People's Liberation Army (SPLA) would not recognize a Sudanese visa. At the same time, as a rebel movement and not an internationally recognized government, the SPLA was not allowed to stamp visas into passports, forcing the would-be traveler to "New Sudan" (as the SPLA liked to call its territory) to go to the SPLA headquarters in Nairobi, Kenya, to obtain a special travel permit—which takes the form of a blue card with your photograph on it. Arriving for the first time in sprawling Nairobi, a city of 3 million people, I asked every taxi driver in sight how to find the Sudan People's Liberation Army, but to no avail. After two days of frustration and a fortune in taxi fares, I decided, on a whim, to look in the Nairobi phone book. And there it was—probably the only rebel movement in the world with an ad in the yellow pages.

Travel permit secured, there remained the issue of the flight. Many of the humanitarian-relief agencies providing emergency aid to southern Sudan were once happy to let journalists hitch rides on bi-planes flying in and out of the Kenyan border-town of Lokichoggio. But by 2005, the only option was an $820 weekly charter flight from Nairobi to Rumbek, the provisional capital of "New Sudan." It boarded in the early-morning hours from an unmarked gate at Nairobi's Jomo Kenyatta International Airport, with no boarding an-nouncement. I tried following the few other people I had spotted carrying blue SPLA travel permits, but quickly lost track of them and ended up on the tarmac wandering from plane to plane in an "Are you my mother?" routine.

The planes touch down in Rumbek on a rubble-strewn dirt airstrip where cattle and goats wander freely. Upon landing, our plane blew out its left tire on a rock and, after disembarking, we all stood in the shade of the wing admiring the gash of melted rubber. An SPLA soldier stamped my blue card and welcomed me with a grin to New Sudan. It was the most hassle-free arrivals procedure I'd encountered in Africa, but looking around this "provisional capital," I realized what it meant to be welcomed to New Sudan. Children were playing in the rusted carcass of a plane that had missed the runway years ago and now had grass growing out of it. Leather-faced cattle herders stood staring at the latest planeload of foreigners to be dropped out of the sky. And next to the runway, a sprawling tent village known as AfEx housed the armies of relief workers, medics, landmine clearers, development experts, and UN officials who had come to "rebuild" southern Sudan.

AfEx was the closest thing Rumbek had to a hotel, so I made my way over and checked in. I was assigned one of the olive green tents with a cot and a towel and told there would be three hot meals a day. A laminated letter on my cot welcomed me to Rumbek, which, it reminded me, "was a battlefield—mines and UXO [unexploded ordnance] are present—stick to well-used paths and roads." It also told me to "avoid walking alone and always carry a handset. There is a risk (as anywhere) of harassment from drunk people and there is a wide availability of small arms." Finally it instructed: "If there is a shooting near the compound: at night—get down on the floor and check your handset for info; during daytime—get down on the ground and crawl for cover."

I didn't have a handset and, being the only person at the camp without affiliation, it seemed unlikely that anyone was going to issue me one. But, despite the slightly *soufflé* security warnings, the place seemed welcoming enough, and I soon forgot all about the handset. In the open-air canteen, doctors and landmine experts stared into the blue glow of their laptop screens, earnestly discussing their "mandates." In the middle of the camp, a round straw-roofed open-air bar sold Ugandan beers and condoms. To the side of the camp, a canvas tarpaulin had been used to create some shade over a television that was hooked up to a satellite dish. Beside that, next to the fence that abutted the airstrip, three local men were laying cement for what I was told would be a second TV lounge. Apparently, AfEx residents had recently gotten into a scuffle over whether to watch rugby or soccer, and the time had come for drastic measures.

"Oh, and by the way, you can't go into town right now," the camp manager told me. "There's been an incident." A driver for the UN Development Program, a member of one of the area's Equatorian tribes, had accidentally run over and killed a much-loved senior SPLA commander. Like most SPLA soldiers, the dead man had been a Dinka, and Rumbek's Equatorian population had gone into hiding, expecting a spate of revenge killings. Police had taken the driver into custody—mostly for his own protection—but in the middle of the night, a mob of Dinkas had broken into the town prison and beaten him to death. Given that a UN vehicle had touched off the incident, the mood in town was judged to be unsympathetic to foreigners, and everyone at AfEx was being told to lie low inside the protected camp for at least a day or two.

But I had not flown halfway across Africa to spend a week in Nairobi chasing down an SPLA permit and running up over $4,000 in travel expenses only to watch South African and Kenyan construction workers argue about whether to watch Sky Sports or the Pope's funeral. The next morning I laid my hands on a motorcycle (there being no cars or taxis in southern Sudan, other than the Land Cruisers used by the UN and other agencies) and ventured into town.

Calling Rumbek a "town" is generous. It is a derelict, haphazard collection of small, round *tikkuls*—rudimentary huts made of sticks and dried grass. Goats lounge in the middle of unpaved roads and barefoot farmers herd bony cattle past the rusted-out remains of armored personnel carriers. Young men, many wearing little more than rags, sit idly with their assault rifles in the dusty soccer pitch the SPLA calls "Freedom Square." Literacy stands at 10 percent and the Diocese of Rumbek estimates that one woman in nine dies in childbirth. Medical services would be nonexistent without the UN and the German relief agency Malteser, and there is no electricity or piped water. In all of the Bahr-el-Ghazal region—an area of some 50,000 square miles—there are only twelve drinking wells. My motorcycle was the noisiest thing for miles.

Most top SPLA officials had disappeared to Khartoum or South Africa for final status talks with the government of Sudan, but it was still a sensitive moment in which to bring up the subject of oil. White Nile Petroleum's controversial deal with the SPLM for the 67,000-square-kilometer drilling concession already claimed by multinational Total had had recently been announced. Total's management and the Khartoum government were incandescent, insisting that the White Nile deal was a sophomoric joke, but the spike in White Nile's shares on London's Alternative Investment Market showed some investors were taking it seriously. Embarrassing headlines were spreading like bushfires through the world's business pages.

Real questions existed about the deal, besides the fact that the acreage was already claimed. For starters, any foreign firm hoping to set up shop in southern Sudan would have to contend not only with the difficult operating environment, but also with ongoing tensions between ethnic groups. As the swift, remorseless slaughter of the hapless UNDP driver in Rumbek had demonstrated so vividly, a bloody "south-south" conflict, with the majority Dinkas pitted against their myriad rivals, was a very real possibility—that is, if the Ugandan Lord's Resistance Army rebels, who were using the area for incursions into Northern Uganda, didn't destabilize the area first.

Whether White Nile would be able to conduct exploration-and-production operations with its meager resources also remained to be seen. Its own technical advisers, Exploration Consultants Limited, estimated that extracting 150,000 barrels of oil a day from the block in question would cost $120 million—no small trick for a company with $15 million in cash and no hard assets.

Then there was the issue of transporting the crude hundreds of miles across landlocked southern Sudan and onto the global market. The SPLM* had made the construction of a pipeline to the Kenyan port of Mombasa a strategic priority for the next six years—the interim period after which, according to the peace agreement signed with Khartoum in January 2005, the people of south Sudan will vote on an independence referendum. Sudan's existing pipeline—a 900-mile duct running straight from the prolific southern oilfields to the northern city of Port Sudan—is a thorn in the side of the SPLM, which has long argued for its right to full stewardship of south Sudan's oil.

For its part, White Nile has batted away suggestions that it is in over its head. When I reached the company's spokesman by telephone

*Sudan People's Liberation Movement, the political wing of the SPLA.

from Nairobi, he gushed about what he called the "White Nile concept," in which the host government gets a 50 percent stake in the company (unusual in Africa, where multinationals generally pay host governments only a small percentage of profits after costs have been recouped). Apparently, I was expected to believe that, unlike big bad Total, White Nile was really in it for the poor, starving Africans. When I got to Rumbek, many senior SPLM leaders I spoke to were quick to voice their skepticism. "No private company would invest anywhere without getting a profit," said Commander Marco Maciec, the SPLM's director-general of political affairs. "For them to talk about fifty percent—let's see it."

In fact, even the most battle-hardened war veterans told me they were troubled by the talk of big oil contracts and easy money. "We are all worried," said Commander Paul Michue, the former SPLA commissioner of Rumbek County, who now spends his days playing cards under a large mango tree. "If you don't have the right kind of management, the right level of accountability and transparency, that worries people who have been at war for a long time."

◄◄—►►

Probably the only people not worried are the Chinese. For them, instability and lack of infrastructure are temporary hurdles on the way to bringing energy security to their country. And it is this long-term approach—this ability to buckle down and simply endure—that Western multinationals find most alien and most threatening about the Chinese presence on the African oil patch. Sudan, in particular, has proven to be a graveyard for Western oil companies. Chevron and Total were forced to abandon their lucrative concessions in southern Sudan in the early 1980s as the north-south civil war broke out, and have been unable to return since. In the case of the American companies, this has been due to sanctions the Clinton administration imposed in 1997. (Though, with southern independence in 2011 a possibility, and a rapidly growing Chinese presence in the area, the Bush administration has found time in its appointment book for the SPLA. Deputy Secretary of State Robert Zoellick made an appearance in Rumbek just two weeks after I was there, his plane greeted by a brass band and cheering villagers.) The Sudanese government's tendency to bomb villages in the south and arm proxy militias has made other Western companies steer clear of the potential PR headache. In 2002 the Canadian independent Talisman—a rare exception to this rule—was forced to give up its exploration program in southern Sudan following a campaign by Canadian human-rights activists alleging, among other things, that the company had provided the Sudanese military with airstrips and other tactical support in its aerial-bombardment campaigns against southern civilians.

But Chinese companies face neither activist shareholders nor independent pressure groups. A Chinese company built the existing 900-mile pipeline and was accused of complicity with aerial bombardment campaigns clearly designed to eliminate southern villages in order to make room for it. Between 1998 and 2000 thousands of people were displaced, and the pattern was always the same. Sudanese Air Force Antonovs would first bombard the villages, and, according to locals, the "Chinese people" would come with bulldozers, followed by government soldiers burning huts. Now as many as 4,000 plainclothes Chinese soldiers are believed to be stationed along the pipeline to protect it from raiders and saboteurs. Spending some $3 billion on infrastructure projects since 1999, China has also built Sudan a refinery and a port, Friendship Hall and Friendship Hospital in Khartoum, a bridge over the River Nile, a rice farm, and a textile mill.

For Western politicians and policymakers, China's growing profile in the African oil business is more than just a commercial threat to Western businesses. In particular, Beijing's growing reliance on African oil has put it on a collision course with U.S. political priorities for the continent. A growing chorus of voices in Washington—from congressmen to newspaper commentators—has been complaining about China's willingness to do business in countries the United States is trying to pressure or isolate. The example most frequently cited is Sudan, whose (notionally) Islamist government many hawks in Washington would dearly love to see destabilized or overthrown, but Beijing's cooperation with Equatorial Guinea and with Robert Mugabe's Zimbabwe is also frequently in the crosshairs.

China relies on Sudan for nearly 10 percent of its imported oil and has invested considerable money and manpower in the Sudanese oil industry. The United States has repeatedly tried to bring a resolution in the UN Security Council against Khartoum for its complicity in what the United States calls "genocide" in the western region of Darfur, but China has made it clear that it would veto any such resolution. Beijing regards Darfur as an internal matter for the Sudanese government to resolve with the help of the African Union—a position as much a reflection of China's traditional live-and-let-live approach to foreign policy, as it is a product of commercial considerations.

Many analysts have cautioned against getting swept up by the rhetoric of yellow peril that has been coming out of certain quarters in Washington, and insist there is a kinder, gentler side to Chinese involvement in African oil that is often overlooked. China has forgiven literally billions of dollars of bilateral debt from African countries in recent years—something Western debt-relief campaigners have tried for years to get their governments to do. It has also set up scholarships allowing some 10,000 African students to be educated in China, and sent hundreds of doctors and teachers to the continent. In 2005 China even agreed to build that desperately needed road between Brazzaville and Pointe-Noire, something no amount of Western aid has ever accomplished.

China's ability to turn relatively small amounts of cash into tangible results, along with its strict regard for state sovereignty, are, for Western governments who still prefer to attach painful conditions to almost any interaction they have with African states, perhaps the most galling aspects of the rising strength of the Middle

Kingdom in Africa. For much of the 1990s Western politicians and economists spoke smugly of the emerging Washington Consensus—the increasingly irresistible belief that trade liberalization, privatization, and free-market economics, rather than oceans of aid money, were a panacea for developing countries. However, China's economic miracle, achieved entirely in the context of old-fashioned top-down state control, has cast serious doubt on this view, and increasingly, in the developing world, one hears wry references to the "Beijing Consensus"—the idea that states should deal with one another as business partners, and then be left alone to manage their own affairs.

Moreover, those who fear China's rapidly strengthening position in Africa might do well to maintain a sense of perspective. After all, the reality is that China has a long way to go before it catches up to the Western presence on the African oil scene. When it comes to exploration licenses, Chinese companies still make do with what one analyst calls "the absolute dregs" and, overall, China's overseas-drilling portfolio is very much in its infancy. Ninety-five percent of the proven reserves of both CNPC and CNOOC* are still inside China. Compare that with the British supermajor BP, for whom the UK accounts for only 7 percent of its reserves, or the three biggest American companies, where the corresponding figures average around 30 percent. For the moment, at least, China's oil industry is most heavily focused on its domestic-drilling program.

It is also worth remembering that China's long-term approach to African exploration carries with it a significant potential for cooperation with the United States. For example, both countries would benefit from greater security in the waters of the Gulf of Guinea. For EUCOM, bringing together the Gulf countries to cooperate on maritime security has been a pet project, and it is easy to imagine a day

---

*China National Offshore Oil Company, the third-largest of the country's NOCs, after CNPC and Sinopec.

when China is invited into the effort. Moreover, Chinese checkbook diplomacy may well backfire. "The logic of the Chinese model may not be sustainable in the long term," says Alex Vines, head of the Africa program at Chatham House, who compares it to the approach adopted by the French state-controlled company Elf, which, until the 1990s, was part and parcel of French foreign policy in Africa. "The French ultimately found the model not to be competitive."

Perhaps more important, much of the hysteria in Washington over China's expanding presence in African oil politics has failed to notice that it is part of a wider Asian search for energy security that simply happens to be playing out on the African continent. The Malaysian state company Petronas, for example, is active in fourteen African countries, including a project with the Chinese in Sudan. In Chad, as part of the ExxonMobil consortium, Petronas has been learning a lot about how to manage a big project, and in coming years will surely become a major player in Africa.

South Korea, meanwhile, has an economy every bit as buoyant and as oil-dependent as China's, with the country now ranking as the world's fourth-largest oil importer. In 2006 the state-owned KNOC picked up valuable new offshore acreage in Nigeria as well as an interest in a block of the Nigeria-São Tomé JDZ. In March 2006 Korean President Roh Moo-hyun, recognizing the importance of Chinese-style petro-diplomacy, visited some of the continent's major oil-producing states, and announced that Korea would be investing $6 billion in Nigerian infrastructure projects, including a pair of power plants that will supply 20 percent of Nigeria's electricity by 2010. It was the first African tour by a Korean president in more than twenty years, and a good deal more successful than the one in 1982 when then-president Chun Doo-hwan had landed in Gabon only to hear the band strike up the *North* Korean national anthem.

But China's most important Asian rival for African oil is India, which is no less desperate to fuel its exploding economy. In 2010 there will be thirty-six times as many cars in India as there were in 1990 and the country's daily oil consumption is expected to rise from the current 2.2 million barrels a day to 5.3 million. Delhi has also made energy security a top priority, spending $1 billion a year in exploration efforts around the world, most of them channeled through the state-owned ONGC.

However, India's approach to securing African oil concessions has been noticeably more timid than China's. India, a rambunctious and cumbersome mega-democracy, makes its decisions slowly, carefully, and transparently. China, a tightly controlled communist state, moves in swiftly and decisively, with little inclination to explain its actions. Again and again in Africa, the Indians have found themselves at the competitive disadvantage. When Angola's Block 18 became available in late 2004, ONGC made what it thought was a convincing bid and, like all the other companies in the running, sat back and waited to hear the results—only to be stunned and infuriated when the block went to China after Beijing's last-minute offer of a $2 billion loan to fund infrastructure projects.

A similar upset happened again in Nigeria a few months later, when ONGC came up with only a 25 percent stake in two prime offshore blocks, despite having made the highest bid. When the results of the licensing round were announced, it emerged that Korea's KNOC had received a 65 percent operating stake after promising to build a gas pipeline for the Nigerians, as well as a shipyard, rail link, and power plant.

Over the course of 2005, the Indians realized their approach to African oil exploration, which some analysts have described as "gentlemanly" and others as dithering or naïve, would have to change. At the World Petroleum Congress in Johannesburg in September, India's petroleum secretary, S. C. Tripathi, expressed his country's bitterness at the way in which African countries were tying the award of exploration concessions to guarantees of cash and development projects.

"Both Nigeria and Angola have conveyed that preference will be given to those offering economic packages," he complained undiplomatically. "How much share you get in a block, they say, depends on the economic-development package you give." A few weeks later, India announced it would be making up to $1 billion available for oil-for-infrastructure deals in African countries. The countries covered by the so-called Team 9 initiative included Chad, Equatorial Guinea, and Ivory Coast, but the ministry has since said it would also target São Tomé and Congo Brazzaville. Much like China, India built up strong friendships in Africa during the heyday of the Non-Aligned Movement. But when it comes to oil, it's all about money, and it is virtually always the case that China has deeper pockets.

And it is not only the Asians who are descending on Africa. Brazil's mammoth NOC Petrobras is perhaps better placed than anyone to lend its offshore and deepwater drilling expertise to Africa, and the Brazilians have been happy to exploit their cultural and linguistic ties with Angola, São Tomé, Guinea-Bissau, and Mozambique—all current or potential oil-producers—in creating what could soon be a powerful global Lusophone oil industry.

Last but not least, there are the Africans themselves. The continent's budding superpower, South Africa, was slow out of the starting gates, but has been catching up quickly. President Thabo Mbeki has made securing exploration deals in Africa a greater priority, and his ANC government has begun making use of the potent pan-African credibility it built over decades of struggle against apartheid, offering South African know-how to African countries such as Nigeria that are keen to "indigenize" their oil industries. In 2005 Mbeki traveled to Khartoum and came back with a lucrative deal for the South African NOC PetroSA to drill in northern Sudan. African pride, always a potent force in South Africa, may also be behind Pretoria's drive to source more of its energy needs on the continent, which it views very much as its own backyard. "It's always possible," says Patrick Smith of the influential British newsletter *Africa Confi-*

dential, "that the Africanists in Mbeki's government are beginning to ask, 'Why are we buying our oil from Iran?'"

American interests are not the only ones to feel the heat from the Asian invasion. Though France has lately put its foreign-policy focus on building a strong, unified European Union, Paris is especially nervous about its declining influence in Africa. To some extent, this declining influence owes a lot to the aggressive new presence of both the Americans and the Asians, but it is also France's increasing aversion to risk in both political and commercial relations with Africa after the embarrassment of the Elf legacy. "All the old Elf managers have gone," says Philippe Vasset, editor of the Paris-based *African Energy Intelligence.* "They are buying virtually no new acreage. They drilled one well in Equatorial Guinea and left, which turned out to be a mistake. They haven't even bid in São Tomé. Total is the one company that knows most about the JDZ at this early stage, because its Nigeria offshore block is right next to the JDZ. But they still aren't bidding. They exited Chad early, because they thought it was too risky. And in the two biggest plays in Africa in the near future—Nigeria and Angola—the French are up against severe competition because it's not their turf. Nigeria is in the U.S.-UK sphere and they can't count on Francophone clout and, in Angola, they suffer from the Falcone legacy."

Total's timidity and political baggage in Africa is mirrored at the level of French officialdom. The French government feels somewhat burned by its assertive interventions in such countries as Ivory Coast and by its strong backing for African tyrants over the years, and is now paying the price by having to sit out and assume a lower profile. "Ten years ago, it would have been clear what France would have done in this current environment," says Olly Owen, West Africa analyst for London-based Global Insight. But today Paris watches from the sidelines as other countries benefit from Africa's oil boom.

The growing international competition for Africa's oil wealth is complex, boisterous, and extremely fluid. Ten years ago, no one could have expected that Libya's Muammar Qaddafi would cozy up to the

United States and Britain, or that Libya's oil would become a hot property for Western oil companies. Nor did anyone expect that Africa's links to its former European colonial metropoles—cultivated over decades and in some cases centuries—would appear so trite and on the verge of irrelevance in the face of Chinese and American economic muscle and realpolitik. Few would have expected Australian, Irish, Swiss, and American independents to be jostling for space with the massive national oil companies of Brazil, India, and Malaysia. Even fewer would have expected South Africa—still blinking into the daylight after years of international isolation—to take a lead role in promoting African energy to the world. But it has all happened, and with such speed that it is hard not to talk in terms of a new scramble for Africa.

From Uganda to Liberia, from Eritrea to Madagascar, and from Sierra Leone to Namibia, virtually every African president hoping to distract from his own domestic failures or internal conflicts has been commissioning seismic surveys, setting up "road shows" in London and Houston, and announcing licensing rounds in the hopes of attracting sizable signature bonuses into his country's threadbare national treasury. Some of these efforts have been based on more dubious prospects than others. In 2004 the Gambia's president, Yahya Jammeh, plagued by a currency crisis and a badly stagnating economy, announced to an incredulous public that the country was sitting on reserves of oil greater than those of Kuwait, and that it was up to every Gambian to pitch in and do his duty in turning the country into an economic powerhouse. In fact, it is unlikely that there is much oil, if any, in the Gambia, though this has not stopped an obscure Canadian company called Buried Hill Energy from signing an agreement with the Gambian authorities.

Elsewhere on the continent, however, efforts have occasionally paid off. In February 2006, for example, the impoverished northwest African state of Mauritania began producing 33,000 barrels a day, thanks to plucky Australian independent Woodside Energy. It may be too early to tell whether the East African margin or the Great Rift Valley will be the "next Gulf of Guinea," as some suggest, but it is clear that interest is beginning to build. In 2006 Aussie independent Hardman Resources began drilling along the shores of Uganda's Lake Albert, an area Uganda's energy minister has called "underexplored." (Shell drilled the last hole there—in 1938.) And it now appears that Uganda is on its way to joining the ranks of African oil producers, with 60,000 barrels a day expected in coming years. The Chinese are exploring Ethiopia, and ExxonMobil has begun exploration in Madagascar. It has been lost on no one that East Africa is positioned for easy transportation links to China via the Indian Ocean, in much the same way that the Gulf of Guinea once was for Europe and North America.

As Vasset puts it, "There's still a lot to do in Africa. In Nigeria, there has been virtually no exploration in deepwater. In São Tomé there's been virtually no exploration. In Equatorial Guinea, the ultra-deep is still unexplored. There is Northern Chad. There is Darfur. There is still a lot up for grabs. The Chinese have bought not just new exploration acreage, but also acreage that is already *producing*. That's what so remarkable about Africa, that you can still do that. It's still a very fluid environment."

Epilogue

IT TOOK ONLY TWO DAYS in Nigeria before I was ready to try peppered snails.

I had read about this national delicacy, but needed to see it with my own eyes. If you're accustomed to French-style escargots the size of a thimble, Nigerian snails come as a big surprise. They are the size of a human fist, and are generally chopped into bite-sized pieces and served in a fiery red-pepper sauce that leaves you drenched in several layers of fresh sweat on top of the layer you've been carrying around all day.

On this particular evening, I was dousing the flames with cold beer in the company of a few other journalists at my hotel in Lagos, when a couple of young Nigerians one of them knew joined our table. They were both about my age, had lived most of their lives in England, and had come back to Nigeria to visit family over Christmas and New Year's. As someone who has spent most of his life divided between countries and cultures, I identified instinctively with their sense of negotiated identity and the frustration of being unable to translate new perspectives into old landscapes. They ranted, as I had expected they would, about how Nigeria would "never change" and how no one wants to listen to someone who's spent time abroad and has a little wisdom to impart.

The conversation turned to oil, which unleashed the usual animated discussion on the rack and ruin oil wealth has brought to Nigeria—the grab-it-while-you-can mentality, the endless military coups, the tribal, ethnic, and religious bitterness, and the paralysis of the state on numerous occasions. "You know, we have a saying in Nigeria," one of them offered in his slick South London accent. "Every man for himself."

I nodded politely and decided not to remind him that this was a British expression, and one he would have picked up in Peckham rather than in Lagos. But then I realized he wasn't finished. "Every man for himself," he repeated. "And God for us all."

I should have known there would be an African version. And I should have known it would involve God. The African continent is a deeply and reflexively religious place, in a way that those of us who have grown up in the moral relativity of the West take a long time to appreciate fully. By the end of my time in Africa, I had lost count of the times I had been asked my religion as small talk, in the same way that Americans will break the ice by asking what you do for a living. The answer people expected was either "Muslim" or "Christian," not, as I would generally reply, that I was "not really a religious man." Even to the most educated and worldly of people I met, that a man might live without God in his life was a mystifying and eccentric concept, rather like answering a question about how many brothers and sisters you have by saying, "I'm not really a family person."

It seemed vaguely reassuring to know that if, in fact, oil was a curse for Africa, and only brought out the worst in people, there was still hope for salvation at the end of the line here, in a way there might not have been in my own godless countries. You get used to thinking in these kinds of absolute terms after a little while in Africa—a realization that first hit me on a warm October evening in Luanda, Angola.

If you take a position high in the hills, and look down, you will see that Luanda is beautiful. The city's concrete tower blocks and pastel-colored government ministries splay out before you in a long arc that hugs the ocean. It's called the Marginal, and it is watched over in turn by the posh embassy district called—what else?—Miramar. And every evening, a watery tropical sunshine will break through the steaming clouds and tickle the whole scene with a final, gasping arpeggio of light, just before the Man Upstairs flips a switch and everything goes black as black as black. Funny just how nonexistent the concept of "twilight" is in the tropics, I thought to myself one night. Not for them the Nordic nuance, not for them the neurotic desire to have everything both ways. No, it is always either night or day in Africa. And you are always either rich or poor.

And then it hit me: I was doing it. I was falling prey to that ancient northern fetish about Africa, the one that had begun with the old mapmakers who wrote "Here there be dragons" all over the Dark Continent, but that continued to this day in the tug-of-war between those who wanted to see in Africa's oil boom the dawning of a new gold rush, and those who saw nothing but misery and theft and plunder, and the curse of oil. Black. And white.

"Africa is about bones and meat and blood and flies," a South African friend wrote to me in an e-mail when I was at one of my lowest points, stuck in the Congo with an expired visa and a looming case of prickly heat, and fast running out of cash. "Everyone is hungry for something." In a way, she was right—but only if you chose to look at things that way. Because if there is one thing I'm sure of now, it's that between the "scramble for African oil" and the "paradox of plenty," there is bound to be a happy medium. Between night and day, there is a twilight in the tropics, though if you blink you may miss it. Between heaven and hell, there is an Africa few of us ever know. And between the red roasting sun that cracks up its soil and the Devil's excrement that bubbles up from below, there might just be—if we learn to look for it—a god for us all.

Acknowledgments

The real heroes of this book already know who they are. For the most part, they either cannot, will not, or should not be named. They are the shadow people, full of courage and generosity. They are people whose names I never caught, or whose lives and careers might be at risk if they were named. Thank you. *Merci. Obrigado.*

As for everyone else, the list of thank-yous could easily turn into another book. Nevertheless, I must express my gratitude first and foremost to my editor at Harcourt, Rebecca Saletan, for taking a gamble on an unknown entity, as well as for razor-sharp feedback during moments of crisis. Stacia Decker also did yeoman's work on the manuscript, performing heroic acts of brush clearing. My thanks also go to my agent, Kathy Anderson, for a skillful piece of matchmaking, and for putting up with my electronic tantrums from halfway around the world.

It was Mitch Albert in London who first nudged me into thinking I "had a book in me." Lubricated with a few rounds of Macallan, I blurted out my big idea about little Equatorial Guinea. Then there was no going back. Here's to your persistence, Mitch; hope you approve.

Olly Owen and Chris Melville are owed a special debt of gratitude. Not only did they provide hours of jovial company in London and believe all my stories, but they also kindly agreed to read a draft of the manuscript and saved me from myself. Here are two men whose encyclopedic knowledge of African politics is matched only by their boundless thirst (for more knowledge, of course).

This project was born in London, so there are inevitably a few more Londoners to thank, beginning with Helen Ferguson for love and support in the early stages, and her parents Ruth and Ant for their kind hospitality. Stryker McGuire at *Newsweek* gave me my first real break into the business and helped nurse all my wacky ideas into print. Madeleine Lewis has been a bottomless source of support and inspiration, not to mention a home away from home on later visits to London. More than anyone else, though, it was the Muses, in the form of Canna Grindley, who dropped in out of nowhere, opened me up like a book, and made me want to write from the heart again. A belated thank-you, my love.

Once it crossed the Atlantic, the book benefited from a timely nudge in the commercial direction, thanks to the nous of Alex van Buren and others. In the later stages, as I plunged into a cavern of debt, a rapidly dwindling circle of loyal New York friends put up with my impending insanity and the increasingly baroque requirements of the "severe austerity plan." Jacob Appel, Natasha Wimmer, Tarik Hussein, Chris Weil, and Danielle Veith in particular suffered this impecunious fool gladly. Tony Charuvastra took the author photo—a fitting capstone to a long and durable friendship. And Joanna Detz, as always, was a source of humor and perspective—a calming, grounding influence, as well as a pro bono graphic design consultant.

Then there was Africa. For special acts of kindness and hospitality, I must thank Michael Peel in Lagos and Lee Grindley in Johannesburg, who let me stay in their homes without being asked—and probably against their better judgment. It took a small army of good souls to get me out of Equatorial Guinea with mind and body intact, and most can't be named. But Mick Hoyle was a godsend, and the delightful Mimi Léonce made me wish I didn't have to leave in the first place.

For their time and wisdom, and for favors great and small—throughout Africa, Europe, and the United States—I am also grateful to Yemisi Adebayo-Payne; Tutu Alicante; Chelsea Bakken and the

U.S. Consular Office in Brazzaville; Luis and Bibi Beirão; John Bennett; Andrea Bohnstedt; Laura Boudreau; the Provincial Government of Cabinda; Mason Colby; João Conde; Bettina Corallo; Trenton Daniel; Oronto Douglas; Adwoa Edun; "Elvis" and his motor scooter in São Tomé; Gen. Carlton Fulford (Ret.); Ian Gary; Sousa Jamba; Calabar's own Peter Jenkins and his troupe of mandrills, who helped me safely navigate the Bakassi Peninsula; Walter Kansteiner III; Chief Bill Knight; Jean-Silvio Koumba; the Ijaw Kingdom of Kula; Lt. Col. Karen Kwiatkowski (Ret.); Brice Mackosso (since imprisoned); the brave youth of Mpalabanda; Nadji Nelambaye; Phil Nelo and the staff of the United States Embassy in Luanda; Rui Neumann; Chris Newsom; Sam Olukoya; Annkio Opurum-Briggs; Roger Bouka Owoko; Tim and Louise Parsons; the "not inconsiderable" Boo Prince; Sanusi Lamido Sanusi; Estelle Shirbon; Jacob Silberberg; Nitza Sola-Rotger; the Sudan People's Liberation Army; Chief Tamro of Ngalaba; Felix Tuodolo; Rev. David Ugolor; Philippe Vasset; Alex Vines; Lodewijk Werre; Sarah Wykes; and Mohamed Yahya.

Sticking up for Big Oil isn't always an easy job, but Joseph Obari and the public relations staff at Shell in Nigeria, Mary Dwyer of Total, Andy Norman and Fernando Paiva of Chevron, and Susan Reeves of ExxonMobil all put their best feet forward and also have earned my gratitude. Shell in particular was kind enough to let me fly around the Delta in its helicopters. It is not a cheap form of transport and the privilege is duly noted.

There is an African proverb that goes something like this: "If today we become soldiers and fight, it is so tomorrow our children will have the peace of mind to become doctors and engineers and politicians, so that their children, in turn, will have the luxury to become writers and dancers and architects." My life has been a luxury, by any standard, and the largest part of my gratitude will always be reserved for my mother and father, and for the battles they have fought.

A Note on Sources
and Suggested Further Reading

This book is based chiefly on hundreds of interviews, background conversations, discussions, and briefings with people throughout Africa, Europe, and the United States between 2004 and 2006. Conversations were generally conducted in English, French, or Portuguese, and any translations into English are my own. For the most part, where these conversations are cited directly in the text, I have identified my interlocutors and provided general dates and locations. In a few special cases, it has been necessary to conceal the exact identities of my sources—either at their direct request, or out of my own desire to avoid exposing them to adverse consequences.

Written sources on which I relied are far too numerous to list exhaustively. They include not only books but also NGO reports; hundreds of articles from trade journals, newspapers, and magazines; Internet journalism; scholarly articles; corporate publications; e-mail correspondence; briefing papers; analysts' reports; official publications from international bodies; and so on. Some of the more important are listed below. I have grouped them thematically with the hope that readers interested in specific issues raised or countries discussed in the text may find this section a useful resource for further reading. In the interest of space, I have generally avoided including newspaper and magazine articles.

AFRICA AND OIL IN GENERAL

Catholic Relief Services. "Bottom of the Barrel: Africa's Oil Boom and the Poor." Baltimore, June 2003.

Center for Strategic and International Studies, Task Force on Rising U.S. Energy Stakes in Africa. "A Strategic U.S. Approach to Governance and Security in the Gulf of Guinea." Washington, July 2005.

Congressional Black Caucus Foundation. "Breaking the Oil Syndrome: Responsible Hydrocarbon Development in West Africa." Washington, July 2005.

Council on Foreign Relations. "More than Humanitarianism: A Strategic U.S. Approach Toward Africa." Washington, February 2006.

Fage, J.D. *A History of Africa*. London: Routledge, 2001 (fourth edition).

Hyne, Norman J. *Nontechnical Guide to Petroleum Geology, Exploration, Drilling and Production*. Tulsa: PennWell Corporation, 1995.

Oliver, Roland, and Anthony Atmore. *Africa Since 1800*. Cambridge: Cambridge University Press, 2005 (fifth edition).

Pakenham, Thomas. *The Scramble for Africa: White Man's Conquest of the Dark Continent from 1876 to 1912*. New York: Random House, 1991.

PFC Energy. "West African Petroleum Sector: Oil Value Forecast and Distribution." Washington, February 2003.

REVENUE TRANSPARENCY

Center for Strategic and International Studies, Task Force on Rising U.S. Energy Stakes in Africa. "Promoting Transparency in the African Oil Sector." Washington, March 2004.

Global Witness. "A Crude Awakening: The Role of the Oil and Banking Industries in Angola's Conflict." London, December 1999.

Global Witness. "Time for Transparency: Coming Clean on Oil, Mining and Gas Revenues." London: March 2004.

Human Rights Watch. "Some Transparency, No Accountability: The Use of Oil Revenues in Angola and Its Impact on Human Rights." New York, January 2004.

NIGERIA AND THE NIGER DELTA

Amnesty International. "Nigeria, Ten Years On: Injustice and Violence Haunt the Oil Delta." London, November 2005.

Falola, Toyin. *The History of Nigeria*. Westport: Greenwood Press, 1999.

Human Rights Watch. "Rivers and Blood: Guns, Oil and Power in Nigeria's Rivers State." Briefing paper. New York, February 2005.

International Crisis Group. "Fuelling the Niger Delta Crisis." Africa Report No. 118. Brussels, September 2006.

Maier, Karl. *This House Has Fallen: Nigeria in Crisis.* Boulder: Westview Press, 2000.

Okonta, Ike, and Oronto Douglas. *Where Vultures Feast: Shell, Human Rights, and Oil.* London: Verso, 2003.

WAC Global Services. "Peace and Security in the Niger Delta: Conflict Expert Group Baseline Report." Working paper for Shell Petroleum Development Corporation, Nigeria. December 2003.

GABON, DUTCH DISEASE, AND THE CURSE OF OIL

Beblawi, Hazem, and Giacomo Luciani, eds. *The Rentier State.* London: Croom Helm, 1997.

Karl, Terry Lynn. *The Paradox of Plenty: Oil Booms and Petro-states.* Berkeley: University of California Press, 1997.

Mahdavy, Hossein. "Patterns and Problems of Economic Development in Rentier States: The Case of Iran," in M.A. Cook, ed., *Studies in the Economic History of the Middle East.* Oxford: Oxford University Press, 1970.

Yates, Douglas A. *The Rentier State in Africa: Oil Rent Dependency and Neocolonialism in the Republic of Gabon.* Trenton: Africa World Press, 1996.

CONGO-BRAZZAVILLE AND ELF IN AFRICA

Catholic Relief Services, Caritas Congo, and Secours Catholique. "Post-Conflict Communities at Risk: the Continuing Crisis in Congo's Department of Pool." November 2004.

International Federation for Human Rights. "Gestion de la Rente Pétrolière au Congo-Brazzaville: Mal Gouvernance et Violations des Droits de l'Homme." Paris, May 2004.

Koula, Yitzhak. *La Démocratie Congolaise Brûlée au Pétrole.* Paris: L'Harmattan, 1999.

Le Floch-Prigent, Loïk. *Affaire Elf, Affaire d'État: Entretiens avec Éric De-couty.* Paris: Gallimard, 2001.

ANGOLA AND CABINDA

Global Witness. "All the President's Men: The Devastating Story of Oil and Banking in Angola's Privatised War." London, March 2002.

Hodges, Tony. *Angola: Anatomy of an Oil State.* Lysaker: Fridtjof Nansen Institute, 2004.

Human Rights Watch. "Angola: Between War and Peace in Cabinda." Briefing paper. New York, December 2004.

Meijer, Guus, issue editor. "From Military Peace to Social Justice? The Angolan Peace Process." *Accord,* Issue 15. London: Conciliation Resources, 2004.

Royal Institute of International Affairs. "Angola: Drivers of Change." London, April 2005.

EQUATORIAL GUINEA

Campbell, Duncan. "Marketing the New 'Dogs of War.'" Washington: The Center for Public Integrity, 2002.

Fegley, Randall. *Equatorial Guinea: An African Tragedy.* Bern: Peter Lang, 1990.

Klitgaard, Robert. *Tropical Gangsters: One Man's Experience with Development and Decadence in Deepest Africa.* New York: Basic Books, 1991.

Liniger-Goumaz, Max. *Small Is Not Always Beautiful: The Story of Equatorial Guinea.* Lanham: Rowman & Littlefield, 1988.

Nze Nfumu, Agustín. *Macías: Verdugo o Víctima?* Madrid: Herrero y Asociados, 2004.

Roberts, Adam. *The Wonga Coup: Guns, Thugs, and a Ruthless Determination to Create Mayhem in an Oil-rich Corner of Africa.* New York: Public Affairs, 2006.

SÃO TOMÉ AND PRÍNCIPE

Frynas, Jedrzej George, Geoffrey Wood, and Ricardo M.S. Soares de

Oliveira. "Business and Politics in São Tomé e Príncipe: From Cocoa Monoculture to Petro-state." *African Affairs*, vol.102 (2003), pp.51–80.

Hodges, Tony, and Newitt, Malyn. *São Tomé and Príncipe: From Plantation Colony to Microstate*. Boulder: Westview Press, 1988.

Mata, Inocência. *A Suave Pátria: Reflexões Político-culturais Sobre a Sociedade São-tomense*. Lisbon: Edições Colibri, 2004.

Seibert, Gerhard. *Comrades, Clients and Cousins: Colonialism, Socialism and Democratization in São Tomé and Príncipe*. Leiden: Brill, 2006.

CHAD

Amnesty International. "Contracting Out of Human Rights: The Chad-Cameroon Pipeline Project." London, September 2005.

Catholic Relief Services and Bank Information Center. "Chad's Oil: Miracle or Mirage? Following the Money in Africa's Newest Petro-state." Baltimore, February 2005.

Human Rights Watch. "The Victims of Hissène Habré Still Awaiting Justice." New York, July 2005.

International Crisis Group. "Chad: Back Towards War?" Africa Report No. 111. Brussels, June 2006.

Petry, Martin, Naygotimti Bambé, and Mireille Liebermann. *Le Pétrole du Tchad: Rêve ou Cauchemar pour les Populations?* Paris: Karthala, February 2005.

SUDAN

de Waal, Alex, ed. *Islamism and Its Enemies in the Horn of Africa*. London: Hurst, 1988.

Index

Abacha, Sani (pres. of Nigeria), 63, 82, 191

Abraham, Spencer, 181

Abubakar, Atiku, 66

Achebe, Chinua, 22

Afghanistan, 139

Afolayan, Samuel, 237

Africa, Sub-Saharan: Asian oil companies in, 205, 290; Chinese building programs in, 277, 287; dependence on oil exports, 14; EUCOM's ambitions in, 233–39; security concerns, 11; growing presence of China in, 205, 273, 275–80, 286–90, 294; historical concepts of, 245–46; humanitarian concerns, 13, 160, 287; importance of religion, 297; oil and "resource curse," 14, 82, 94–98; oil supplies, 9, 12; potential oil boom, 166–67; small independent oil companies in, 218–19, 221–24

Africa Confidential, 292–93

Africa Energy Intelligence, 176, 230, 293

African Oil Policy Initiative Group (AOPIG), 8, 89–90, 92, 183

Agip (Italian oil co.): in Nigeria, 169

Air São Tomé and Príncipe, 241

Alamieyesegha, Diepreye: and oil revenues, 65–66

Alcala, Gen. Raoul Henri, 233

Alfonso, Juan Pablo Pérez: on "curse of oil," 96

Algeria, 5, 166, 277

al-Qa'ida: Saddam Hussein supposedly linked to, 89–90, 93

Amerada Hess (U.S. oil co.): in Equatorial Guinea, 182–83; in Gulf of Guinea, 170, 174, 177

Amin, Idi (pres. of Uganda), 171

Amnesty International, 26

Anadarko Petroleum (U.S. oil co.): in São Tomé, 231

Anaximander, D. Y.: political philosophy, 91

Ancram, Michael, 192

Anderson, Brian, 28

Anglo-Iranian Oil Company, 3

Angola: apartheid system, 164; becomes U.S. ally, 152–54; bilateral trade with China, 276; Catholic Church and social activism in, 162–63, 165; Chinese reconstruction loans to, 138, 148–52, 276, 291; civil war, 127–29, 133–37, 142, 146, 163, 223, 226; competes with Nigeria, 232; Cuban military intervention in, 127, 159; culture of corruption, 116, 141; "curse of oil," 135–36, 167; demand for peripheral oil industries, 130; extractive resources, 128; fascist Portuguese government controls, 163–65; government embezzlement of oil revenues, 135–36, 138–42, 163, 167; government reactions, 140, 147;

Made in the USA
Lexington, KY
22 June 2012